Understanding Genetics

Understanding Genetics

E. B. FORD

PICA PRESS

New York

Published in the United States of America in 1979 by
PICA PRESS
Distributed by Universe Books
381 Park Avenue South, New York, N.Y. 10016

© E. B. Ford, 1979

Library of Congress Catalog Card Number: 79-63132
ISBN 0-87663-728-4

Printed in Great Britain

To the Hon. Miriam Rothschild
in recognition of her outstanding achievements in
the study of insect ecology and mimicry

CONTENTS

PREFACE

Genetics is the study of organic inheritance and its more immediate consequences. It is a young science. Its basic laws were made known to the world in 1900; though an account of them had been published, and ignored, thirty-four years earlier. During that interval, the chromosomes, which carry the hereditary material, had been discovered. It might have been possible, therefore, to make rapid progress in the subject during the first decade of this century. The results did not justify any such hope. From 1910 to about 1920, however, the chromosome mechanism of heredity was being analysed with great success by the associates of T. H. Morgan in the U.S.A. Then, largely as a result of their labours, genetics began to blossom after the First World War. There followed the mathematical approach of R. A. Fisher, together with his theory of dominance, and the revealing analysis of cell division by C. D. Darlington with, much later, his recognition of the supergene. Meanwhile, genetics was being applied to anthropology and medicine and, aided by the polymorphism concept, was making possible the study of evolution in progress. Working on very different lines, J. D. Watson and F. H. Crick elucidated the chemical structure and action of the hereditary units.

Already for a considerable time, therefore, men of general education, and scholars in other fields of learning, have wished to know something of the theoretical aspects of genetics and its practical applications. This book is an attempt to meet their needs.

Any such endeavour will surely satisfy one basic human requirement: the desire to criticize. By some, this work will be held too elementary to be useful; by others, too advanced to be understood. The latter view, even if to some extent valid, would be a fault in the right direction, as it would at least help the most deserving. Indeed experience suggests that the intelligent reading public is better informed and more trouble-taking than critics would have us believe. When Lancelot Hogben wrote *Mathematics for the Million*, he carried non-mathematicians into the calculus, and they loved it. At least the sales of his book suggested they did. He may not actually have catered for a million, but he was certainly read by hundreds of thousands. While I anticipate no such success, nor pose hurdles so severe as he did, I do envisage the possibility that some may be interested enough to wish to advance further than I, in general, dare to take them. Here and there, therefore, I say in effect: those who want can go straight ahead, while for any who like to examine the matter rather more closely, here is a passage designed to meet their needs which, none the less, can be omitted without loss to the main theme. I have added also a brief list of references. So for those who wish to look further, the means will be at hand if they are able to consult a library that caters to some extent for science; and if it be one with open access to the shelves, Utopia is theirs.

Finally, it seems right to clear up a point or two about terminology. I give many of the technical terms of genetics in addition to explaining the ideas they cover, because a knowledge of them will be presumed in works likely to be consulted in further reading. In general I give and use the scientific names of plants and animals as well as the English ones, for these latter are not known, or at all easily discovered, in other countries (p. 59), and it would be wrong to write only for the British public.

In the preparation of this book, as in my career as a scientist, I have received constant help and encouragement from the Hon. Miriam Rothschild. She has carried her own highly original biological concepts into practice, especially in respect of insect mimicry and the chemistry of the substances that protect aposematic species. Her achievements, international in repute, place her among the chief experts in her field of research.

The Nuffield Foundation for long gave financial support to the study of Ecological Genetics at Oxford. The annual cost of our field work on the butterfly *Maniola jurtina*, described in Chapter 3 of this book, as well as on other subjects, is still defrayed by a sum most generously placed at our disposal by the Foundation. Also I, personally, am deeply indebted to All Souls College, Oxford, which has in a variety of ways supported my scientific studies. Furthermore, I am most grateful to Professor J. W. S. Pringle, F.R.S., who has provided me with accommodation and facilities in the Department of Zoology, Oxford.

It is a pleasure to acknowledge much valuable help from Sir Cyril Clarke, F.R.S., Dr. E. R. Creed, Professor W. H. Dowdeswell, Dr. H. B. D. Kettlewell, Dr. E. R. Lees, Professor K. G. McWhirter and the late Professor P. M. Sheppard, F.R.S. I am much indebted to Mr. J. S. Haywood for his assistance in my own researches on subjects dealt with in this book.

                                                                    E. B. F.

1. The Principles of Genetics

The Basis of Heredity

It may seem a platitude to say that the offspring of buttercups, sparrows and human beings are buttercups, sparrows and human beings. But is it so obvious that they should be? I have seen a buttercup with double flowers, also a piebald sparrow, while rarely are two human beings alike. The plant and the two animals vary, that is to say, mankind apparently the most: why so? What then keeps them, and indeed living things in general, 'on the right lines'? Why are there not pairs of sparrows, for instance, that beget robins, or some other species of bird: why indeed birds at all? Something must be handed on from parent to offspring which ensures conformity, not complete but in a high degree, and prevents such extreme departures. What is it, how does it work, what rules does it obey and why does it apparently allow only limited variation? Genetics is the science that endeavours to answer these questions, and much else besides. It is the study of organic inheritance and variation, if we must use more formal language.

Personally, in addition to being a scientist, I happen to be a classical scholar and therefore to some extent a historian. So that when a particular subject is brought to my notice I, in common with many other people, want to know something about the history of thought and endeavour that lies behind it; and in regard to genetics, that history is a curious one. There was already speculation about organic inheritance in the days of ancient Greece; but the relevant *facts*, based upon experiment, only began to emerge in 1865 as the result of thinking in the right way.

In that year, Gregor Mendel (1822–84),* a monk in the Augustinian Monastery of St. Thomas at Brünn, now Brno, in Moravia,

* He had studied mathematics, physics and biology in Vienna, had visited England and was elected Prälat (Abbot) in 1868. There exists a semi-'underground' tradition or story that when Mendel was in England he visited, or may have visited, Darwin; but secretly, because of the almost certain disapproval of his monastery. I am myself the last of those who shared their friends with Darwin, and among the last who knew one of his children (Major Leonard) quite well. I have made enquiries upon lines

gave two lectures to the Natural History Society of that town; they were published in its *Transactions* for 1866. In them he laid down the basic principles of genetics confirmed by experimental breeding work on the edible pea, which he grew in his monastery garden.

Here was a notable step forward in human knowledge. Its immediate result was what the cynic might expect: it was almost completely neglected; until the end of the century, in fact. This has never been satisfactorily explained. The *Transactions* of the Brünn Society are not too easily available, but they were to be had in most of the chief scientific academies of Europe. In England there were sets, of course with Mendel's papers in them, in the libraries of both the Royal and the Linnean Societies.

At any rate, in 1900 three biologists (de Vries, Correns and von Tschermak-Sysenegg) chanced independently upon Mendel's work, and each realized that here was something remarkable. Separately they drew attention to it in current journals, and in a few months Mendel had become world-famous. By then he had been dead for sixteen years. Yet, as he had said himself, 'Meine Zeit wird schon kommen.'*

It will presently become clear what it was that Mendel achieved (pp. 15–23). His results are not in doubt, but the means by which he reached them are uncertain. It is generally supposed that he drew his conclusions from his cleverly designed crosses using different types of edible peas, which he carried out for eight years. Yet behind them there may lie a strange, unexplored background. We now realize that it would at least have been possible to arrive at most, though not all, of his deductions as a pure mental exercise. If he did so, his plant breeding would largely have been by way of confirmation. Perhaps that was what happened; we are not sure, but there is some evidence to suggest it (Fisher, 1936†).

I have already attempted to recapture what may conceivably have been Mendel's line of thought (Ford, 1973) and, reconstructing it in the light of modern knowledge, to carry it rather further than he could have done. It is well worth recapturing, for it lies at the root of the whole matter. Moreover, it will be a surprise to find that so simple, logical and evident a trail had not been followed earlier. It was this failure that induced Charles Darwin to adopt views on heredity incompatible with his own theories of evolution;

---

no longer available, and am confident that no meeting between Darwin and Mendel ever took place.

* 'My time will certainly come.'

† For details of this and subsequent references, see p. 184–6.

and, to this very day, it leads writers on genetics to describe Mendel's work as if no background to it had been at hand.

First, it seems logically inescapable that some sort of *units* responsible in some kind of way for the features of plants and animals must be handed on from parent to offspring if heredity has any physical basis at all. It is convenient to anticipate by calling such units *genes*; though the name was not given to them until 1909.

The next question is this. Are the genes transmitted wholly, or chiefly, by one sex or in approximate equality by both parents? The latter alternative proves to be correct, except for a minor adjustment (pp. 45–9), and there is a sound basis for that decision quite apart from any preconceived views that may be held on the matter. This is obtained by a technique in elementary mathematics, known as 'correlation'. But there is no need to be a mathematician to understand its use. For it is merely a device which shows to what extent the variability of two features is associated. The result is expressed by a number, the 'correlation coefficient',* which can take all values from 0, complete independence, to 1·0 when the association is absolute: as it is for instance between weight and volume in a number of solid spheres of the same material but of different sizes. Generally, of course, the correlation coefficient is fractional, as it would be if the spheres were of different materials. There would then merely be a tendency for the larger to be the heavier.

But, instead of estimating the extent that two qualities are associated in one group of objects, it is as easy to work the other way round and estimate the association of one quality in two groups. If the latter be parents and offspring in plants or animals, a definite correlation shows that the quality is inherited and to what extent.

Thus, in measuring the height of a group of fathers and their adult sons (or daughters), some association is found between the two, for the taller fathers tend to have the taller children. Now the correlation coefficient proves to be effectively similar when calculated for the same group of children and their mothers. That is to say, the hereditary *contribution* of the two parents to the height of their children is equal. (It does not matter that the average height of the mothers is less than that of the fathers; all that is important here is the fact that the variation in the heights of the two parents is equally reflected in that of their children.) Moreover, probability mathematics can establish whether any difference between the two correlation coefficients is a real one, or due merely to chance.

This result has the widest application, and is therefore very rele-

---

* A simple and convenient method of calculating it is given by Clarke (1964, pp. 328–33).

vant to practical affairs. It shows, for instance, how mistaken is the view that the male line is more important than the female in stockbreeding.

Here we must step aside for a moment to think of the microscopic structure of animals and plants. The living substance of the body, the *protoplasm*, is generally enclosed in minute units, the *cells*. Some of these, known as *germ cells*, are set aside to produce the actual units of reproduction, the *gametes*: a useful term, as it applies to either sex. In animals, the female gamete is the *egg* (or *ovum*) and that of the male is the *sperm*. This latter is exceedingly small and swims in fluid produced by glands associated with the genital organs. In plants, the female gametes are the *ovules*, while those of the male are carried in the minute *pollen grains* that form the dust-like pollen.

Since the two sexes contribute in approximate equality to the heredity of their offspring, the simplest possible situation is that the genes are present *in pairs* (known as *alleles*), the members of which are derived respectively from the male and the female parent. Thus, any *one* hereditary quality must be controlled by *two* genes. That indeed is the general situation, though there is sometimes a number of possible alleles from which the pairs may be chosen (p. 17).

Evidently the pairs of genes forming each allele must separate from one another at reproduction, one being transmitted to each offspring; otherwise, the number of genes would increase at each generation. Such separation was predicted by Mendel, and named by him *segregation*. It ensures that each gamete, whether male or female, receives one member only from each pair of alleles. The pairs are restored in the ensuing generation when, in animals, the sperm fuses with the egg to produce a new individual; while a corresponding process occurs in plants. Thus the alleles consist of one paternally and one maternally derived gene.

But it is also important to realize, as Mendel did, that such segregation also allows reassortment (recombination), and therefore inborn variation. The pairs may be brought together again as they were, or combined differently. Consider a gene C, for flower colour, constituted as C^rC^r for red flowers, and C^wC^w for white, in the alleles. Segregation pulls each apart, as C^r and C^r, also as C^w and C^w. When restored at the next generation by crossing, they may reappear as C^rC^r and as C^wC^w as before; or differently, but still in pairs, as C^rC^w.

The next step establishes the fact that the hereditary material retains its identity from generation to generation. That is to say, the genes do not contaminate one another, or blend, as biologists in the past thought they did: a view that was the most devastating mistake that Darwin ever made (p. 14). Thus heredity is said to be *particulate*; and again, there is a general test to establish the truth of that assertion.

Let us think of any variable feature in a plant or animal during three consecutive generations, obtained by taking a pair of individuals and interbreeding their offspring to give rise to inbred grandchildren. The original pair is in breeding experiments conventionally known as the 'first parental generation' (P1), and their sons and daughters (rather odd terms when using plants, but they will serve) as the first filial generation (F1); while the inbred grandchildren of P1 are denoted F2. These latter possess no genes that their parents did not have (except for 'mutation', which is so rare that it can be discounted; see below) since they are the product of brother and sister mating. With genetic blending, the F2 generation must be less variable than the F1, for blending leads to uniformity: that, indeed, is one of the things that is meant by blending. If, on the other hand, the genes retain their identity and can be shuffled about into new combinations at each generation, the F2 individuals will be *more* variable than the F1, and this is always found: it is the very reverse of blending.

Here, of course, we are concerned with the absence of blending between the *genes*. The blending of the features they produce is quite usual (p. 18), but irrelevant in the present enquiry, since it does not affect the hereditary material.

A similar comparison is applicable also to the human race, even though there is an objection to incest in most communities. For it can be obtained, though less simply, by comparing the children of near relations with those of parents who are only remotely related.

It has already been mentioned that each gene controls a given feature, or set of features, in the body, the allelic pairs co-operating in the process. Within that restriction, then, the genes can exist in distinct forms: taking, for instance, one of the alleles for flower colour in a plant, which may perhaps be red or white. Another example is length of fur in a mammal, which may possibly be long or short. Indeed the alternative genes may not be limited to two: there may be a number of them, forming *multiple alleles*, all controlling the same feature or set of features in varying degrees. Any two of these, but not more, can then exist in the same individual (but see pp. 108–9 for exceptions to the latter statement).

One allele can be transformed into another by an exceedingly rare type of change known as a *mutation* (pp. 56–7, 167). This is more frequent in some genes than others, but a mutation rate of one in a million individuals is perhaps a fair average.

Mutation takes place instantaneously, giving rise to a new and very stable gene referred to as a *mutant*: a pair of terms with the same verbal distinction between them as 'conception' and 'concept'. Evidently, then, the members of an allelic pair may be either similar, when they are called *homozygous*, for instance both responsible for red flowers or both for white; or alternatively they may be dissimilar, *heterozygous*, owing to mutation in one of them, perhaps in the remote past: one responsible for red flowers and one for white, a situation that might be expected to produce a pink shade.

It will be asked why these clumsy and unfamiliar terms should be used rather than the easily understood 'pure-bred' and 'hybrid'. The answer is that the latter do not cover the ideas in question, for they refer to the plant or animal as a whole. If two geographical races are crossed, the offspring are correctly said to be hybrids between them; but some of their alleles will be made up of genes in the same state (homozygous) and others of genes in dissimilar states (heterozygous).

It is now possible to think of a simple breeding experiment, using a common European, and British, moth: the Buff Ermine, *Spilosoma lutea*. The gene controlling its wing colour may be referred to as C. This can take the form C^y, a yellowish shade, with some black dots, which is the normal one, and C^b, giving rise to an uncommon blackish variety. Both are homozygotes, C^yC^y and C^bC^b. Each sperm and each egg of the latter type will contain one colour gene C^b; so that on crossing two of the black insects the homozygous alleles C^bC^b will again result and the form breeds true. The same applies to the yellow insects, giving rise to sperms and eggs carrying C^y.

But if the parents, known as the P1 generation, are both homozygotes, yet of different kinds, C^yC^y and C^bC^b, each will contribute one member of its pair to the (F1) offspring. These then are heterozygotes in which the relevant alleles will be reconstituted as C^yC^b. In appearance they are in this instance *intermediates*; black with a yellow patch in the fore wing and yellow lines, like rays, along the nervures (the supporting struts of the wings). The three forms are illustrated in colour in my book *Moths* (1976).

What will happen when the two heterozygotes are crossed? The result can be worked out in a few minutes. The two genes representing the allelic pair in question, C^yC^b, will separate into equal numbers of germ cells as C^y and C^b. On interbreeding to give the F2 generation, the chances are equal that C^y from one parent (say the male, though it could be either sex) meets C^y or C^b from the other, resulting in C^yC^y and C^yC^b alleles in equality. So also for the equally numerous partner C^b, giving C^yC^b and C^bC^b also in equality. Thus we have the situation shown in Fig. 1.1.

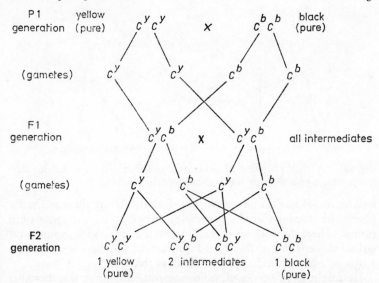

Fig. 1.1. A cross between two different, but allelic, colour forms of the Buff Ermine moth. It shows the production of their F1 offspring, also of the F2 generation in a 1 : 2 : 1 ratio. This includes the original P1 types once more (yellows and blacks), 'extracted' *pure* out of the F1 intermediates.

Evidently then, three types arise, and in a ratio of 1 : 2 : 1. This naturally is the one always obtained when combining at random two objects having alternative forms in equality. Put thus, as it should be, the expression is lumbering and perhaps obscure; but the occurrence is obvious, down-to-earth and commonplace. If two coins are tossed together 100 times, they will fall both heads about 25 times, a head and a tail about 50 times and both tails about 25 times: a ratio of 1 : 2 : 1, as in genetics and for a similar reason.

The diagram also confirms the fundamental fact already stressed: that although the *effects* of the genes do, in this instance, blend when brought together, the genes themselves do not, for they retain their identity in the heterozygotes. Thus the P1 types reappear in F2, having lost nothing of their purity. These 'extracted' yellows and blacks could be used to found unsullied lines of yellow moths and of black.

It is important to notice that a mating between a homozygote and a heterozygote gives rise to homozygotes and heterozygotes *in equality*, as shown in Fig. 1.2. Such a mating is known as a 'back-cross', simply because it can arise when a heterozygote is crossed back to a homozygous parent or, of course, to the parental type.

Mendel himself made crosses between one variety of edible pea

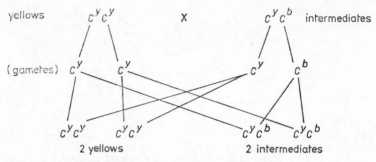

Fig. 1.2. A mating between a homozygote and a heterozygote, showing that this produces homozygotes and heterozygotes in a ratio of 1 : 1.

having round seeds, and one with wrinkled.* Using these as the P1 plants, he obtained an F1 generation in which all the seeds were round. These he interbred, and raised F2 peas with round and wrinkled seeds in a ratio of 3 : 1 (his actual numbers were 5,474 : 1,850 = 2·96 : 1). What did he think of that?

He might have supposed, following something like the thoughts we have followed here, though without the special proofs, that he would get a 1 : 2 : 1 ratio; but he did not. Yet he could hardly be in doubt what was happening; for the F1 plants showed that the heterozygotes were not intermediates, as he no doubt expected, but resembled one of the homozygotes, in fact the round-seeded type. But the potentiality for wrinkled seeds had not been obliterated by any blending of the hereditary material; since they appeared in the F2 generation and, moreover, in the expected proportion of one quarter. It could have been deduced at once that the F2 rounds were made up of homozygotes and of heterozygotes that resembled them, as seen in F1.

However, it must have been a great satisfaction to Mendel when he showed that the F2 rounds were indeed composed of two visibly indistinguishable classes in a proportion of 1 : 2. He had his 1 : 2 : 1 ratio after all.

Actually, although he did not know it, the homozygous and heterozygous rounds can be sorted out visibly into their two types, but this requires the observation of their microscopic structure. The starch grains, seen under the lens, are in the homozygous rounds relatively large, elongated and separate. In the wrinkled form they are smaller, irregular in shape and often compound. The starch grains of the heterozygotes are intermediate in type.

The best way of showing that the F2 rounds belong to two

* The seeds must be allowed to dry before this feature can be distinguished.

classes is to cross them with recessives (wrinkled). The homozygous rounds then produce only rounds, as in F1 in Fig. 1.3 (p. 23); the heterozygotes give rounds and wrinkleds in a 1 : 1 ratio, being a back-cross.

A feature that appears the same in a heterozygote and one of the homozygotes (whether due to one gene or to two, that is to say) Mendel described as *dominant*. He described as *recessive* a feature only detectable in a homozygote, and therefore requiring two genes to make it apparent. We use these terms today. Thus the round seeds of the pea are dominant, while the wrinkled condition is recessive. We meet here for the first time a fundamental point. The terms 'dominant' and 'recessive' apply to the *effects* of the genes (that is, to *characters*, as we say), not to the genes themselves. A gene has multiple effects, some of which may be dominant and some recessive, or those qualities may be absent, in the starch grains of our example, in which two genes have a greater effect than one. To be dominant, the effect must be similar in the homozygotes and heterozygotes; there is no such thing as a 'dominant gene' or a 'recessive gene'.

Mendel worked with six other characters in the edible pea. The most striking related to height: Tall,* 1·5 to 1·8 m, dominant to short, 23 cm to 46 cm in the stock he used. In this instance, no way of distinguishing the heterozygotes from the homozygous dominants is known; and that is by far the more usual situation.

It seems likely that Mendel arrived at the idea of segregation (p. 16) both theoretically and from a study of his results in counting peas. It involved also what may be called the 'purity' of the type: that each gene is uncontaminated by any other, for the F2 homozygotes are pure though their genes have passed through the F1 intermediates.

He next turned his attention to another question, one which had to be answered from his experimental work. When two or more pairs of alleles are segregating at the same time, does the behaviour of one pair influence that of the others? It might be, for example, that those brought into the cross together from the same parent tend to remain together or, perhaps, that the dominants all tend to segregate in a group. The answer he obtained was that no such associations exist, and the segregation of one pair of alleles is independent of any other: this is Mendel's 'rule of independent assortment'.

That is to say, Mendel reached the view that if we consider the occurrence of two 3 : 1 ratios in the same generation, each behaves

* Following the usual practice, capital letters will be used for dominant characters, small ones for recessive.

as if the other were not there: so that when combined, they assort in a ratio of $(3 : 1)^2 = 9, 3, 3, 1$.* This is, of course, what we get when we toss together two pairs of different coins (say, two bronze twopenny pieces and two 'silver' tenpenny pieces in the new British coinage) calling 'heads' dominant and 'tails' recessive. Anyone who cares to practice this will notice how the result gets 'better' (approaching $9 : 3 : 3 : 1$ more closely) as he increases considerably the number of throws (say, 25, 50, 100). Of course perfection may be obtained with very few throws, not often.

Let us satisfy ourselves on the point by using the two qualities in the pea already mentioned: Tall plants dominant to short, and Round seeds dominant to wrinkled. It is a great help to use the same letter for a particular gene in its different forms so that we can trace the behaviour of the two segregating pairs without confusion. For this purpose, in the same way as with characters, a capital letter is used for the gene controlling the dominant feature we are studying and a small letter for that giving rise to the recessive character.† We may therefore use S for the gene giving rise to tall peas and s for that producing short ones; also W for the gene for Round seeds w for the wrinkled.

We can therefore start our cross with the two different double homozygotes: $SSWW$ (Tall with Round seeds) and $ssww$ (short with wrinkled ones), this being the P1 generation. The members of the pairs of alleles segregate from each other to form on the one hand SW gametes and on the other sw. On crossing to give the F1 generation, only double heterozygotes, $SsWw$, producing the two dominant characters, can be formed. All the plants of that generation are therefore alike (Fig. 1.3). What will happen when these alleles segregate and themselves form gametes? The chances are equal that S is combined with W or w to form SW, Sw; so too with the equally numerous s gene. Four gametic types consequently arise in equality: SW, Sw, sW, sw. Such genes brought in by the F1 gametes can be combined in sixteen ways in F2, as shown.

An inspection of the composite F2 square gives further information at a glance. The four plants along the diagonal passing downwards

* Without dominance, one pair of alleles segregates, as we know (p. 19), in a ratio of $1 : 2 : 1$; so that two pairs produce $(1 : 2 : 1)^2$ forms. That is nine classes, distributed as $1 : 1 : 2 : 2 : 4 : 2 : 2 : 1 : 1$.

† This illogical system dates back to the early days of genetics and is generally retained only because so entrenched that it would cause confusion to abandon it. It is, however, quite unreasonable, in view of the fact that one gene may produce several features, some dominant and some recessive. The use of capitals and small letters for *characters* is quite reasonable, except that dominance and recessiveness can vary (pp. 100–1).

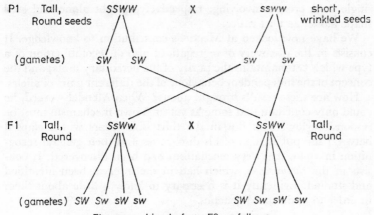

P1　Tall,　SSWW　　　　X　　　　ssww　short,
Round seeds　　　　　　　　　　　　　wrinkled seeds

(gametes)　SW　SW　　　　　　　sw　sw

F1　Tall,　SsWw　　　X　　　SsWw　Tall,
Round　　　　　　　　　　　　　　Round

(gametes)　SW Sw sW sw　　　　　SW Sw sW sw

These combine to form F2 as follows :

		SW	Sw	sW	sw
	SW	SSWW Tall Round	SSWw Tall Round	SsWW Tall Round	SsWw Tall Round
F2	Sw	SSWw Tall Round	SSww Tall wrinkled	SsWw Tall Round	Ssww Tall wrinkled
	sW	SsWW Tall Round	SsWw Tall Round	ssWW short Round	ssWw short Round
	sw	SsWw Tall Round	Ssww Tall wrinkled	ssWw short Round	ssww short wrinkled

Result :　Tall, Round seeds, 9 ;　Tall, wrinkled, 3 ;　short, Round, 3 ;　short,
wrinkled, 1.

Fig. 1.3.　Mendel's rule of independent assortment. Here we see the effect of crossing plants homozygous for different phases of two pairs of alleles, dominance being complete for each. It will be noticed that Tall and short growth, considered alone, segregate as 3 : 1 in the F2 generation, as does seed shape. It will also be found that in the absence of dominance nine genetic types are produced, as indicated in the footnote on p. 22.

to the right are the four possible double homozygotes; those along the other diagonal are the double heterozygotes. The lowest row gives the double back-cross result, now four types instead of two, but still in equality, as 1 : 1 : 1 : 1. The two middle rows show the single back-crosses involving, respectively, height alone and seed shape alone, as 1 : 1 ratios.

We have now looked at Mendel's contribution to knowledge. It consists in the discovery of segregation; also of recombination of a type which yet maintains the purity of the hereditary units; and the concept of the independent behaviour of the different pairs of alleles.

How are these results brought about? When Mendel worked, he could only conclude that some as yet unknown mechanism must be responsible for them. But in the thirty-four years which elapsed between the publication of his discoveries and their general recognition in 1900, that very mechanism had been discovered. It consists in the *chromosomes*, which had in the interval been identified and studied. Evidently it is necessary to know a little about them in order to understand genetics.

The Chromosomes and Linkage

In considering the mechanism of heredity, our attention is surely to be focused on the cells, in which the protoplasm (p. 16) is generally housed. Now the gametes, the reproductive units themselves, that is to say, are cells also, but of a special type. They, and nearly all others, consist of two parts: a small *nucleus*, enclosed in a membrane, which controls the rest of the cell protoplasm, known as *cytoplasm*. The amount of this latter, but not the size of the nucleus, varies from one cell to another; it does so strikingly between the gametes of the two sexes, for generally it is immensely greater in those produced by the female. Indeed large eggs, of birds and reptiles for example, may contain a million times as much cytoplasm as the sperm, yet the hereditary contribution of the two sexes is the same. Clearly, therefore, the genes must be restricted mainly to the nucleus. There they are located in minute rod-like or thread-like structures, the *chromosomes*. Although these persist, they can only be seen (microscopically) and studied in detail when the cells divide. Their number is generally constant in a species, and can vary from a single pair to many dozens.

After the 'rediscovery' of Mendel's findings, it was soon realized (by W. S. Sutton, in 1902) that the chromosomes must carry and distribute the genes. For the behaviour of the latter, as shown by breeding experiments, corresponds exactly with that of the chromosomes as seen under the microscope. Let us then compare them.

The chromosomes, as the genes, are present in all the normal body cells in pairs, the members of which are derived respectively from the two parents. This is achieved by the accurate process of cell division, known as *mitosis*. It reduplicates one cell accurately into two products, dividing each pair of chromosomes and genes.

Yet in the germ cells the members of the pairs of chromosomes (*homologous chromosomes*), and of genes (the *alleles*), separate from one another, one member of every pair passing into each of the gametes. These therefore contain the half number of chromosomes, and of genes, found in the body cells. The total (double) number is restored when a sperm fuses with an egg at fertilization. The chromosomes are present in the gametes in what is known as the *haploid number* (n) and in the body-cells as the *diploid number* (2n) (but see pp. 108–9). They can be counted under the microscope; while the total number of genes in any organism is, of course, unknown, though it must be halved in the gametes. There is, then, a step-by-step correspondence here.

Can it possibly be doubted, therefore, that the chromosomes carry the genes and provide the mechanism for segregation and recombination? Obviously not; yet as we shall see, Bateson, who was the most distinguished figure in genetics at the beginning of this century, did doubt it (p. 35). It may be added that the perfect correspondence between rare chromosome abnormalities and abnormal genetic segregation has now supplied precise proofs of the chromosome theory of heredity.

Since the genes are responsible for the hereditary qualities of the organism, their number must be very large: thousands, compared with a few dozen chromosomes. Therefore each chromosome must carry many genes. This, as realized almost at the beginning of the century by Sutton, must produce exceptions to Mendel's rule of independent assortment which, though applicable to genes in different chromosomes, does not appear to be so to those in the same one. The tendency for genes to assort together, instead of independently, because they are carried in the same chromosome, is known as *linkage*. When the genetics of a species have been sufficiently studied, it will be found that the genes which show linkage with one another fall into several groups, the number of which is the same as the number of pairs of chromosomes seen under the microscope.

It might be thought that no reassortment is possible between *linked genes*, those that are carried together in one chromosome. A little knowledge of chromosome behaviour alters that point of view. It could indeed be predicted that some method of breaking down linkage must exist; otherwise the chromosome mechanism of heredity would be intolerable: it would mean that useful genes could not be separated from harmful ones when linked with them.

In the normal cells of the body, each chromosome is split longitudinally into two parts, the *chromatids* (that is to say, the future chromosomes), in preparation for the next cell division, but long in advance of it. Now the last two divisions of the germ cells before gamete formation are of a quite exceptional kind. They are known as *meiosis* (first and second): and during the two stages the nucleus divides twice but the chromosomes once only: therefore their number is halved. When the first meiosis occurs, the split forming the chromatids is long delayed, the homologous chromosomes being brought together, and so held, by an attraction between the pairs of alleles, exactly facing each other, which they carry: that being an essential requirement of gamete formation. At a late stage in meiosis one therefore finds sets of four thread-like bodies lying together. These consist of the two homologous chromosomes, each now split into two chromatids. If asked how many such sets of four there will be, surely the answer is easily given: it will be the haploid number of the organism (in man, for instance, twenty-three). When the split forming the chromatids does at length take place, a small piece, the *attachment constriction*, is not at first affected by it. This, then, still unites the two chromatids comprising each chromosome.

Moreover, the pairs of homologous chromosomes are also held together, but by very different means: by an interchange of one or more blocks of material, and therefore of genes, between one chromatid from each (Fig. 1.4). Such an interchange is known as a *chiasma*; and it will be noticed that two out of the four chromatids are unaffected by it. Evidently it allows a transference of linked genes, known as '*crossing-over*', between homologous chromosomes: a breakdown in linkage.

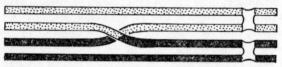

Fig. 1.4. *Chiasma formation.* A single chiasma has formed between a pair of homologous chromosomes (represented, respectively, black and stippled). These are each split into a pair of chromatids which are still united at the spindle attachment, here shown near one end. It will be seen that the chiasma involves two chromatids derived from different homologous chromosomes. These have interchanged material but not partners at a point, while the other two have remained intact. (Reproduced from Ford, 1942, by kind permission of Associated Book Publishers Ltd.)

If two pairs of genes are located in different pairs of chromosomes, there will be free assortment between them, producing four back-cross classes in equality, as in Fig. 1.3. If the two pairs of genes are carried in the same chromosome and there is no crossing-over between them, only two back-cross classes can arise: they are the two parental ones. But when crossing-over (due to a chiasma) occurs between the two pairs of genes, an interchange between them takes place; it will allow some members of the two recombination classes to appear, in equality. Their frequency is indicated by the percentage of these two recombinants added together. This is known as the *cross-over value* (*C.O.V.*). The situation is easily understood from an example.

In the maize plant ('corn' in the USA), Coloured seeds (*G*) are dominant to colourless (*c*), the Round ones (*R*) are dominant to wrinkled (*r*). We can think of the back-cross between the double heterozygotes *CcRr* (being Coloured, with Round seeds) and the double recessives *ccrr* (colourless and wrinkled). If these pairs of genes were on different pairs of chromosomes, we should obtain four classes in equality (p. 23): two being the parental types (*CcRr*, *ccrr*) and two being the recombinations between them (*Ccrr*, *ccRr*). In this instance they are, however, in the same chromosomes, so that if no crossing-over occurred between them, only the two parental classes could appear among the back-cross progeny. Actually, a small amount of such crossing-over allows a few of the recombination classes to arise:

	parental classes		recombination classes	
characters	Coloured, Round	colourless, wrinkled	Coloured, wrinkled	colourless, Round
genes	*CcRr*	*ccrr*	*Ccrr*	*ccRr*
class numbers (per cent)	48·2	48·2	1·8	1·8

If, however, the two parental types had been *Ccrr* and *ccRr*, it is they that would have been in excess in the back-cross (at 48·2 per cent each) while the recombination classes would have been *CcRr* and *ccrr* (at 1·8 per cent each). That is to say, there is no tendency for one allele to assort with another as such; whichever go into the cross together tend to come out together, and to the same extent. Consequently, we here calculate the cross-over value (C.O.V.) as 3·6: a result obtained whichever classes happen to be the recombination ones.

What the C.O.V. may be is a matter that depends on the amount of interchange and the position of the genes on the same chromosome. That is to say, the chances are greater that a cross-over shall occur between two genes if far apart, *D* and *S*, than if they are near together, *S* and *U*:

Thus a relative estimate can be made of the distance apart of the genes on the chromosomes.

It is possible also to determine their arrangement and order by a study of three linked genes; say, *A*, *B*, *C*. It is found that if the cross-over value between *A* and *B* is 2 and that between *B* and *C* is 5, then the cross-over value between *A* and *C* will be either 7 or 3, the sum or the difference. That result can only arise if the genes occupy definite positions, their *loci*, and *in a line* along the chromosomes. Moreover, if the C.O.V. be the sum, their order is *ABC*; if the difference, it is *ACB*. Thus we can find the arrangement of the genes, their order and their relative distances apart. That is to say, we can map the chromosomes.

This is a real step forward, and in that work we are mapping something invisible. Fig. 1.5 shows the chromosomes of the fly *Drosophila melanogaster*,* while Fig. 1.6 is a map of a few of its genes. It will be seen that they fall into as many 'linkage groups' as there are pairs of chromosomes; also that three of the chromosome pairs

Fig. 1.5. The chromosomes of a fly (*Drosophila melanogaster*), greatly enlarged. There are three long pairs and one small, dot-like pair. The pair at the the bottom are the sex chromosomes (Chapter 2), X to the left and Y to right.

are long and one is very short, while three of the linkage groups contain a relatively large number of genes, some far apart, and one a very small number all close together. That the latter are those in the little chromosome is now beyond doubt. When a rare abnormality has affected it and it alone, the small linkage group but not the others has shown abnormalities also.

Finally, one additional point may be mentioned here. Since any one cross-over involves only two of the four chromatids belonging

* In 1968 the estimated total of the genes so far studied in that insect was 992, and it would be considerably more today; while the chromosomes amount to four pairs.

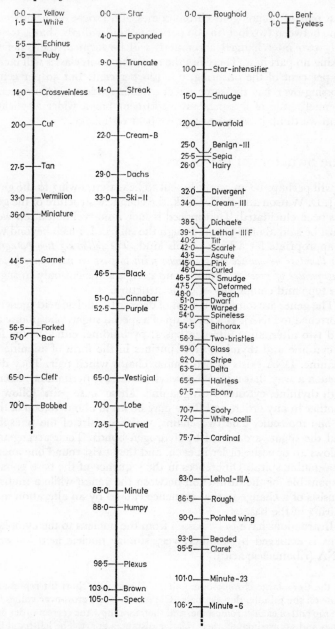

Fig. 1.6. The position of a few of the genes that have been mapped on the chromosomes shown in Fig. 1.4. There are three long lists of genes, being those

to a pair of homologous chromosomes, if a cross-over occurs every time between two loci (in 100 per cent of individuals, that is to say), only two interchanged chromatids will be formed, the other two taking no part in it. Therefore the recombination classes will include 50 per cent of the offspring, not 100 per cent. But 50 per cent of crossing-over has the same effect as independence. Consequently, Mendel's rule of independent assortment has a wider application than we think it has when first we hear of linkage.

The Structure of the Genes

It will perhaps be an astonishment to hear that, owing to the genius of J. D. Watson and F. H. Crick, the chemical structure of the genes has been elucidated. That indeed is such a marvel that a few words must be said about it, even though the subject is much beyond what is appropriate for a book of this kind. *The following two paragraphs can therefore be omitted by those who wish to pass on more directly to the consequences of genetics.* What is said here is sufficient only to suggest the type and complexity of genic structure.

The genes are found to consist of deoxyribonucleic acid (generally shortened to DNA). It is made up of a special sugar, phosphoric acid and two alternative nitrogen bases: pyrimidine, either in the form of cytosine and thymine, or of purines in the form of adenine and guanine. DNA exists as two similar chains which pair. They do so in such a way that a purine always partners a pyrimidine: adenine with thymine, cytosine with guanine. These 'base pairs' follow one another in any order, and there may be up to about 10,000 of them in one molecule. The two chains, which consist of the phosphate and the sugar, are united by hydrogen bonds. Their arrangement follows an opposite order in each, and they twist round one another in a double spiral. Differences in the sequence of the base pairs are responsible for the distinction between the genes, while a mutation consists of a change in that sequence caused by an alteration in the pairing of the bases.

Instructions have to be passed from the nucleus to the cytoplasm. This is achieved by another, very similar nucleic acid known as RNA (ribonucleic acid).

---

on the three large chromosome pairs, and one very short list representing those on the minute, dot-like pair. The numbers are cross-over values from the top end of each chromosome, and therefore show the correct order of the genes and approximately their relative distances apart. The left-hand list is of sex-linked genes (Chapter 2) carried in the X chromosome (Crew, 1925, p. 117, Fig. 32).

The chemical substitutions that a gene can undergo appear to be restricted. It is this that enables us to say that the alleles control similar but contrasted characters.

Environmental Variation

The genes interact with one another and with the environment to produce the effects for which they are responsible. It is possible to compare the importance of genetic and environmental variation in a number of ways: by direct experiment; mathematically by means of correlation (the degree to which features are associated between parent and offspring); and, in man, by a study and comparison of twins. These, of course, are not very common in the population; but they are of two kinds, fraternal and identical. The fraternal type is derived from the fertilization by two different sperms of two eggs produced at the same time. They can be similar or dissimilar in sex, and they differ genetically as much as ordinary brothers and sisters born singly. Being of the same age, however, they tend to be brought up in a more similar environment than usual, and this slightly increases their resemblances. However, about one-third of all twin pairs are identical: that is to say, they arise from a single fertilized egg, one that splits into two at a very early stage in development. They therefore are of the same sex *and have the same set of genes*. In consequence, any differences between them, which indeed must be slight, are purely environmental (p. 183). In general, they are exceedingly alike. We have all encountered these astonishingly similar twin pairs, two boys or two girls, and there is rarely any doubt about their distinction from the more usual fraternal type. At need, the matter can be settled from a study of their blood groups (pp. 155–6), for these will almost always be the same if the two children are identical (for the possibility of very rare exceptions, see Clarke, 1964, p. 284). The tendency to produce twins in a family, of one kind or the other, is inherited.

On the average, about half the variation of a plant or animal is genetic, and half is environmental. The impact of the environment upon the body can, however, differ greatly in dissimilar conditions and between the sexes. Thus in the butterfly *Maniola jurtina*, many genes contribute to the production of spotting on the underside of the hind wings (pp. 60, 63). At 15°C about 14 per cent of the variation in spot number is genetic in the male, and 63 per cent is so in the female. At 22°C the values become 47 per cent for the male, and 78 per cent for the female.

There are two extreme alternatives by which plants and animals

B

may adjust themselves to their environment, and between them every gradation may be found. On the one hand, they may respond rather accurately to the diverse conditions they encounter. In a little shrimp-like animal, *Gammarus chevreuxi*,* an inhabitant of brackish water, the rate at which black pigment is deposited in the eye, when red, is closely associated with increasing temperature. This type of adjustment advances stepwise with the environment.

On the other hand, plants and animals may adjust to a type best suited to the whole range of conditions that they normally meet. This situation can well be illustrated by the work of Kettlewell on the Scarlet Tiger moth, *Panaxia dominula*. He identified a gene in it which can produce a highly abnormal colour pattern, less extreme in the heterozygotes than in the homozygotes (an effect I have illustrated: Ford, 1976 *b*, Plate 14). However, these variants cannot be detected except when the moth is reared at a constant temperature. Thus the abnormal response, doubtless disadvantageous, has been obliterated except in an environment that the insect never encounters in the wild. Consequently a given pattern is here adapted to all the conditions that the moth can meet.

Though environmental variation can be studied in mammals, including man, these animals carry their own constant temperature about with them; and this, of course, is that at which they are adjusted to live. Here, then, is an important aspect of their surroundings to which they do not in general respond. However, there are a few instances in which the critical point for the action of a gene is so close to the mammalian constant temperature that slight departures from it affect the genic action. For instance, the Siamese cat, with all the features that characterize it, is nearly recessive to the normal type. Its unusual pale coloration is one of its most obvious qualities. This is due to the gene in question acting at the ordinary, and accurately maintained, body temperature. Yet with a drop of a few degrees it has an entirely different effect, giving rise to black fur. There is no sign of that colouring at birth, since the kittens have been kept warm in the maternal uterus. Subsequently, however, the extremities, the tip of the nose, the ears, paws and tip of the tail, are sufficiently exposed to fall slightly below the general temperature of the body; and consequently they darken. That this is not a direct pattern effect can be shown by shaving a patch on the back. The exposed skin is thus cooled, so that the fur is black when it starts to grow again; but it becomes paler once more when long enough to raise the warmth of the shaven area. A similar, but

* Actually, though shrimps and *Gammarus* are both Crustacea, they are far removed from one another within that group.

still more striking, effect is found in the 'Himalayan pattern' of the rabbit, for this has white fur with black extremities.

Some genes act, or their effects are intensified, at high temperatures (as Bar eye in the fly *Drosophila melanogaster*) while for others this is true only for low ones. For example, a form of recessive albinism in barley is not total unless the plant has been grown below 6·5°C; while above 18°C no action of the gene can be detected.

Moreover, in plants environmental variation may be a somewhat complex reaction to the qualities of the soil, including its acid or alkaline state, as in *Hydrangea*. Environmental effects can indeed widely influence the features of plants and animals. Two further types may suffice by way of example. A gene in *Drosophila melanogaster* obliterates, more or less completely, the regular banding of the abdomen, provided that the larvae are reared in moist conditions. If the cultures are allowed to dry, that abnormality no longer occurs. As is generally known, the distinction between queens and workers in the hive bee is produced by feeding.

The appearance of some animals differs widely at one time of year from another: that is to say, they are subject to seasonal variation; and this must be the result of the environment. The butterfly *Araschnia levana*, which is widespread in western Europe, including France, has two generations in the year. On the upper side it is reddish brown, marked with black bands and dots, in the spring; while in the summer the wings are blackish, with a somewhat interrupted white band crossing them (Ford, 1971, Plate 35). The difference, which gives the effect of highly distinct species, is produced by temperature acting during the first twenty-four hours after pupation.

There are of course various types of interaction between environment and heredity; examples of the kind are not infrequent in human diseases, among other conditions (*see* p. 163). Rickets is due to a deficiency of vitamin D, which can arise from an insufficient amount of those substances in the diet which supply it, such as eggs, butter, milk and cheese. It can be due also to underexposure to sunshine, for ultra-violet light allows this vitamin to be manufactured from certain precursors in the skin.

Rickets is a common condition among children of the poorer classes brought up in slums in large towns. Yet it arises much more easily in some constitutions than in others: the tendency to respond to certain environments by developing rickets is indeed partly genetic, and is due chiefly to a single gene with heterozygous expression. Consequently the disease is common in some families, but rare in others, even when all are exposed to similar conditions.

It will now be evident that genes cannot be equated with the effects (the 'characters') they may produce. It has already been said that there is no such thing as a dominant or a recessive gene; but neither is there a gene for this or that bodily condition (not even a 'black-eyed gene', as I have seen in the literature!). For the genes have multiple, and often widely different, effects, and what they do in one environment they may not do in another. Two useful terms may therefore be introduced at this point. A *genotype* is an organism judged by its genetic constitution; a *phenotype* (from the Greek φαίνομαι—*phainomai*: 'I appear') is one judged by its 'characters' in the widest sense: anatomy, appearance, chemistry, physiology, habits and so forth. Evidently, then, the phenotype is the product both of heredity and of the environment.

I remember an earnest lady, a Gladstonian Liberal, who devoted her life to what she considered improving mankind and meliorating the lot of the lower classes, who said to me: 'I hear you study genetics. I wish instead you would study the effects of the environment.' She was no more ignorant than might have been expected; the study of genetics necessarily includes the effects of the environment.

Two Geneticists in the History of Science

At the beginning of this chapter we thought briefly of the history of genetics, so it is appropriate to end it with reference to two geneticists: William Bateson, great in his own day, and Mendel himself, great in posterity.

Early this century, William Bateson (1861–1926) was the outstanding figure in genetics. He named the science; and to think of him throws a flood of light upon the first stages of its development. He was already an exceptional man when attention was drawn to Mendel's work. He had discovered *Balanoglossus*, the most primitive of the Chordata (the group to which the vertebrates belong). He had with extraordinary determination, overcoming great hardships and difficulties, carried out for two years, from 1886, a zoological exploration of the Russian steppes; though he did not obtain the results he had hoped for. Subsequently, he became the first Professor of Genetics in the world (at Cambridge); though he quickly moved on to be the first Director of the now celebrated John Innes Horticultural Institution. It was he who, overtly at least, first showed that Mendel's laws apply to animals as well as to plants, and we owe to him certain terms in general use today: 'allelomorph' (now generally shortened to 'allele'); also 'homozygote' and 'heterozygote', and in seeing the need for the two latter terms he showed his penetration. He was a man who did important things, but he was hardly in the accepted sense a *great* man. Coming into genetics at the start, he could hardly have escaped making a name in it; yet it was percipient of him to realize that here was something not only new but fundamental. He was completely non-mathematical and far from open-minded.

It is astonishing and hardly credible that he did not accept the chromosome theory of heredity and did not, therefore, understand linkage when, in collaboration with Punnett and Saunders, he actually discovered it, though it had already been predicted as a logical necessity by Sutton (p. 24). In order to explain it and yet escape the chromosome concept, he proposed an alternative, 'gametic coupling', which need not detain us as it is wholly to be rejected. Yet he clung to it in spite of accumulating evidence to the contrary. This was obtained principally by Morgan, Bridges, Sturtevant and Muller, at Columbia University, New York, in their researches on the fly *Drosophila melanogaster*. So overwhelming did that evidence become that in 1923 Bateson went to New York to see for himself what was the status of the *Drosophila* studies.

One day that year I happened to be in London and chanced to meet Major Leonard Darwin, son of Charles Darwin, in Piccadilly. He said, 'Bateson is back from America, I have just seen him in the Athenaeum.' 'What did he say about the *Drosophila* researches at Columbia?' I asked. 'He was reading and I did not disturb him,' said Darwin, 'but it would be interesting to know. Let us go back there and get his views.' So Darwin and I went to the Athenaeum, a few minutes' walk away, and found Bateson. After a minute or two of general conversation, Darwin said, 'I believe you have just seen what they are doing with *Drosophila* in New York, so do you believe in the chromosome theory of heredity now?' 'Yes, I do,' said Bateson, 'and all my life's work has gone for nothing.' The remark was unpremeditated and untrue, but it shows how deeply Bateson felt on the subject.

Finally, we can return briefly to Mendel, and the problem which his life and work presents. In the first place, we may merely note how fortunate it was that he did not happen to encounter, or at least recognize, an instance of linkage, or use a polyploid plant (pp. 108–14). For although his results, had he done so, would have been strictly in accord with his own conclusions, they would at that date have appeared to him as flagrant exceptions, which he could not have understood, since the chromosomes had not then been discovered.

More interesting is the fact that his researches face us with a penetrating question on the value of scientific evidence. The total

number of plant and animal species now described lies between
1,100,000 and 1,200,000. It would have been far less in Mendel's
day, but still very large. Yet he based his views upon a single one
of them: the edible pea, *Pisum sativum*, in a number of races and
subspecies. It is true that he corroborated them to a slight extent
by work on, unfortunately, a related plant: the bean, *Phaseolus*. He
also published the results of his experimental crosses with hawk-
weeds, *Hieracium*. These were, in fact, unsatisfactory because that
genus is to a considerable extent apogamous, growing from un-
fertilized seeds, that is to say; a fact not known in Mendel's day.
He also studied bees extensively; but his notes on the subject are
lost. Yet, owing to the reproductive peculiarities of that insect
(p. 78), it is not likely to have provided him with obvious opportunities for confirming his views on heredity.

Thus Mendel's conclusions, though probably developed from a
consideration of living organisms in general (pp. 14–17), were really
only established from his monumental study on peas. Is it surprising,
therefore, that when attention was drawn to them in 1900, many
biologists thought that they threw light upon heredity transmission
in that plant only; or possibly upon that of the Papilionaceae, the
family to which it belongs? Are the principles apparently derived
from experiments upon a single species really applicable to over a
million others, exhibiting all the diversity of animal as well as of
plant life? It seems questionable indeed. Oddly enough, it would
not have done so had Mendel merely used one other, chosen with
discrimination. Had he shown that his analysis of heredity applied
equally well to a higher animal, say a mammal, as to a higher
plant (such, indeed, as the pea), being forms of life near the apex
of the two main evolving lines, his conclusions could at once have
qualified as a great generalization.

Mendel must surely have appreciated this. It is unthinkable that
a man of his intellectual power did not test his results obtained
from botany upon, for example, mice; which he is actually known
to have bred. He could, therefore, at least have raised F2 generations and back-crosses from matings between, say, normally
coloured animals and albinos; a task that presents no difficulty to
children.* Why did he not publish such results and so suggest the
universal application of his theories? We can only guess, but
probably not wide of the mark. It seems likely that in his day and
country the Church would have objected to a monk, and a pro-

---

* In mice, normal colouring (grey) is a simple dominant to the recessive
albino. The results of crosses between the two kinds are, therefore, exactly
the same as those in any of the seven features studied by Mendel in his peas.

minent one (he became abbot of his monastery), conducting breeding experiments on mammals. Thus his career, which had provided him with leisure and opportunities for his investigations, may also have prevented him from demonstrating their fundamental significance.

Even so, the mystery is not fully solved. Had Mendel shown the working of his laws in an invertebrate, such as an insect, in addition to his peas, their wide application would have been established, and no religious difficulty would have arisen. Possibly his unfortunate choice of bees for experimental purposes, proving as it probably did unsatisfactory, did not encourage him to carry breeding work on insects any further.

2. Sex

Sex is directly or indirectly determined by genes responsible for male or female development. Yet the *control* of sex is vested in specific chromosomes in which some of the genes concerned are carried. The ways in which these act will here chiefly be related to the human situation, because it is judged that the educated man who is not a biologist, for whom this book is intended, will be particularly interested in this impact of genetics upon mankind.

Sex Determination

As in the majority of higher animals, males and females are present in something like equality in the human population; and in such a clear-cut fashion that individuals obviously intermediate between them in structure are very rare indeed. Here, then, we have segregation, so it is reasonable to expect that genetics can throw light upon sex determination.

That process is indeed a simple one, so it is extraordinary that the devices for achieving it should be so diverse. Though only one of the two parents actually determines the sex of the offspring, there is no uniformity as to which. Certainly it seems more often the male that does so, as in mammals and therefore man; yet in other groups, the Lepidoptera (butterflies and moths) and birds, for instance, the female lays two kinds of eggs: male-determining and female-determining. Moreover, in fishes quite closely related species differ in this respect.

Yet even within one or the other system, a corresponding result may be obtained by very different means. Now, as far as we know, the human mechanism is somewhat exceptional, so it will be helpful first to lay down the more normal, or perhaps the more basic, plan from which it departs.

As already mentioned, the reproductive material consists of the eggs of the female and the sperm of the male; also, these are in an important sense each equivalent to half a normal cell of the body. For the number of their chromosomes is halved during their formation, but in a special way: such that each egg and each sperm receives one member of every chromosome pair; inevitably, therefore, one member of every pair of alleles. The pairs of both are restored at fertilization, because this consists in the fusion of a sperm with an egg to form the first cell of the new individual.

Now the number of chromosomes possessed by different species is very diverse; as already mentioned, two or three pairs only or many dozens. But a large or a small number does not represent a highly evolved condition which others are striving to attain. The fruit fly *Drosophila melanogaster* has four pairs; man has twenty-three; and the Chalk-hill Blue butterfly, *Lysandra coridon*, has ninety.

One chromosome pair out of the total consists of 'sex chromosomes' which are responsible for the *control* of sex. All the others, the 'normal' chromosomes, are known as 'autosomes'. The two sex chromosomes are alike in one sex, and are known as X chromosomes. In the other they differ: one is an X chromosome also but its partner is of a unique type, the Y chromosome (Fig. 1.4, p. 28). This distinction is not merely one of the features of sex, like the sexual organs, but the formation of the male or female depends upon it. The unlike pair is the one found in the parent deciding the sex of the offspring: it is therefore present in male mammals and female birds (p. 48).

Not only, of course, does each reproductive cell receive *one* member only of every pair of autosomes but also of the sex chromosomes. We may therefore look at the behaviour of the latter; seeing this first in the body cells, then reduced to half in those of the sperm and the egg and finally restored to its full number again by the additive nature of fertilization:

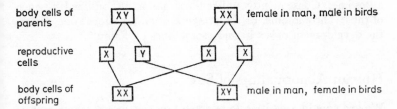

We have here a simple method for producing two clear-cut sexes in equality. How does it work?

Within the distinction of XY being male or female, three methods of sex-control are theoretically possible here. One of them, however, can be eliminated at the outset. This is the situation presupposing different kinds of X chromosomes, male-determining and female-determining. We can exclude it, however, because as a glance at the diagram shows, each X chromosome can be interchanged between the two sexes. But two possibilities, or combinations of them, remain. Either the number (though not the kind) of X, 1 or 2, is decisive, Y being immaterial; or the Y chromosome gives rise to one sex and its absence to the other. The first of these seems to be the basic form. It is that which characterizes the fruit flies, *Drosophila* (XY male) and the Lepidoptera, butterflies and moths (XY female). The second, Y being decisive, is the one principally operating in man.

We must consider this a little further, though briefly. In the first, or more basic type, genes tending to produce males or females, respectively, are scattered among all the chromosomes, but disproportionately. That is to say, where XY is male, the autosomes are predominantly male-promoting and X is predominantly female-promoting. Here the fixed dosage of excess male determinants in the autosomes outweighs the feminizing influence of one X, but is outweighed by that of two. The mechanism works the other way round when XY is female. In such a system, Y is unimportant in sex control. In fact this operates effectively even in abnormal circumstances in which Y is lost, or if by mistake two Y chromosomes (XYY) are present. Yet Y is not inert if it exists at all, and it does not always do so; for such XO males in *Drosophila* are sterile. Evidently in these fruit flies Y carries genes necessary for male fertility, though not for male structure. Obviously it can never carry any that are essential for the working of the body, since the XX sex is without it.

Now we turn to the situation found in man. It is that in which the male-determining genes are principally carried in the Y chromosome, balanced against female-determining ones in X and in the autosomes. Consequently, unlike the situation just outlined, absence of the Y chromosome, or the presence of an additional one, affects the development of sex in man, sometimes profoundly.

Human Abnormalities of Sex

We are now in a position to comment on this matter, some aspects of which are of considerable social consequence. Here we shall deal only with abnormalities related to sex-determination; that some diseases are commoner in one sex than the other is a subject for Chapter 7.

It will be useful to think first of numerical differences in the human sex chromosomes. It sometimes happens that, owing to an error in cell division, the Y chromosome is lost from an ordinary (XY) male, producing XO, and added to the female, which therefore becomes

XXY. It will be realized that this must have a severe effect in the human situation, but not where sex depends upon the number of X chromosomes.

The XXY individuals are women largely converted to men and passing as such. The condition is called Klinefelter's syndrome, and those affected by it are known as Klinefelters. They can often be identified because, though possessing male genital organs, they are somewhat immature and feminine in type: the face is smooth, the voice high and the breasts enlarged. In certain tissues (e.g. the mucous membrane of the mouth) the cells of women possess a microscopic structure, the 'Barr body', absent from those of men, and this is present in Klinefelters. Such people are liable to mental deficiency, though some are fairly, and a few quite, intelligent. Their marriage can be consummated; but they are almost always sterile, since they rarely form sperms. They may seek advice because of their enlarged and sometimes painful breasts and failure to produce children. Yet many must pass unrecognized; indeed it may seem surprising that the armed forces include such women undetected. Evidently their frequency in the population is uncertain, but at least one 'man' in 1,000 is a transformed woman of this kind.

The incorrect cell divisions which produce XXY give rise also to the reciprocal condition of XO. This is the reverse of the Klinefelter, comprising men largely transformed into women, the so-called 'Turner's syndrome'; but it is much more serious. Such people are in appearance stunted, immature girls who hardly ever menstruate; and they frequently suffer from a dangerous abnormality of the main artery of the body (the aorta). Thus, with a tendency to early death, they are on the one hand rarer in the population than their counterpart; yet, on the other, they are nearly always recognized, since their parents seek advice when their supposed daughters fail to have their proper 'periods'. If detected in the first few years of life, the condition can be somewhat improved by administration of the male sex hormone (p. 42).

The exceptional distribution of the Y chromosome, producing Klinefelters and Turner's syndrome, throws further light upon the control of sex itself. For it has been suggested that the human X chromosome is destitute of genes for sex, the balance of which, it is then contended, consists in the fixed dosage of preponderantly female-determining genes in the autosomes acting in the presence or absence of Y. Since, however, the addition of an extra X to the normal male chromosome set (XY to XXY) has a marked and strongly feminizing effect, that alternative view cannot be sustained.

Another form of incorrect cell division affecting the sex chromosomes leads to the presence of an extra Y, that is to XYY. This, as might be expected, causes an intensification of the male attributes. Such men are exceptionally tall: in one sample, their average height was 186 cm whereas that of the corresponding normal group was 170 cm; the difference was highly significant on the numbers available. Moreover, the XYY type tends to be ill adjusted, being aggressive in a way which often leads to crimes of violence, so that such people may find their way into prisons. Here we have an instance of the widely established fact that intelligence and psychology are under genetic control (and see pp. 182–3).

It will have been noticed with surprise how variable are the arrangements leading to the general and, one would have thought, quite simple production of two sharply defined sexes in approximate equality. The matter is subject to still further complications, one of which must be met at this stage. That is to say, in many groups of animals the sex of each part of the body is that of the cells composing it. Therefore, if an incorrect cell division occurs anywhere, it may produce a recognizable patch of male tissue in a female or the reverse. Sometimes this happens at the first cell division after fertilization, which results in a creature that has one half of the body male and the other female. In these circumstances, castration or even replacing the reproductive organs with those of the wrong sex has no evident effect save to sterilize the individual, or to cause it to produce reproductive cells of the incorrect kind.

That does not occur in man or other mammals, for the sex of their body is not the sex of its parts. The XX, XY chromosome mechanism decides whether the ovaries, forming eggs, or the testes, forming sperms, shall arise in development. But in this group of animals these structures do more than that. They produce a hormone* which, circulating in the blood, determines whether the tissues shall develop along male or female lines. In these circumstances, castration evidently has an effect upon the body as a whole.

We now have the information to discuss homosexuality. Though individuals obviously intermediate in structure between men and women are very rare indeed, it will be evident enough that great opportunities exist for minor deviations in sex. For sexuality is based on the condition that the chromosomes, autosomes as well as sex chromosomes, contain many genes tending to force development, mental and physical, in the direction of the inappropriate as well as of the overt sex. These, like any other genes, can mutate and, in consequence, segregate, so as to produce variation in the type and

---

* Hormones are manufactured by glands and passed direct into the blood. They are necessary for the development or working of distant parts of the body.

amount of sex hormones and the response of the body to them. In consequence, some men have in certain respects a feminine bias and some women a masculine one.

Not only are the genes controlling sex numerous, but they can affect different aspects of its many attributes. Consequently, one finds some homosexuals taking a more masculine and active part in homosexual relations, and others a more submissive one.

It has been pointed out that the effects of the genes can be influenced by the environment, genetic (provided by the gene-complex) and external. So it is with sex. There is no doubt that the circumstances of childhood and youth can promote or minimize homosexuality, but in some constitutions rather than in others: lack of opportunity for getting to know the opposite sex; or lack of opportunity to gratify bisexuality, as among merchant seamen and sometimes in the armed forces.

But the absence of such restrictions is by no means decisive in the matter. It is not infrequent to find one member of a large family homosexual when the rest are not, although all may be brought up in apparently similar conditions. Nor is this unexpected. For though it has been said that clear structural intermediates between the sexes are rare in the human species, as in most other animals, homosexuals often tend in physical respects to depart slightly towards the opposite sex. The voice of certain types of adult homosexual men is rather high and of women rather low; so also the characteristics of the body (e.g. the size of the breasts and the distribution of pubic hair) may be slightly intermediate. Such facts confirm that the condition has a physical as well as a psychological component. Consequently, another point must be mentioned here. A basic concept of genetics is the principle that genes can control the time of onset and rate of development of processes in the body (as shown by myself and Huxley, 1927; see also pp. 76–7). It is not surprising, therefore, to find that in sex, which is attained as a balance between masculinizing and feminizing influences, it is in youth even normal for ordinary men and women to pass through a somewhat homosexual phase; and this can become a fixation.

Both Klinefelters and those with Turner's syndrome are below the average standard of intelligence. Not so the homosexuals. Indeed in them the reverse seems true; as if a mind somewhat adjusted to these exceptional emotions were a particularly receptive one. Some of the greatest men of the world have been homosexual (Rowse, 1977).

It is impossible to estimate how frequent homosexuality may be, since the condition cannot be clearly defined. It ranges through every degree from those who take some slight sexual interest in their own sex to the, by no means rare, state in which men or women happily married with children value homosexual friendships, culminating in those to whom a heterosexual liaison is unthinkable. Even if one could take the halfway point onwards in such a series, there is of course much concealment. This works in one direction only, to reduce the apparent proportion of homosexuals below its true value. Certainly 1 percent of all men and women would be too low a proportion of the whole population, and numerically even that represents a vast total.

Considering its genetic aspect, among adults at any rate homosexuality is really incurable, in spite of the claims to the contrary made by psychiatrists. Moreover, it is probably true to say that the great majority of homosexuals have no wish to be cured.

Taking these facts into consideration, it is important that physicians, on the occasions when they are consulted in the matter, perhaps by parents of homosexuals, should not give the impression that the condition is a shameful one or that it should be considered criminal. The latter point needs to be emphasized in view of an English law, due to Labouchère, passed in 1885 but repealed in 1967. This made homosexual practices illegal between men but not between women, a distinction of almost incredible stupidity. That law applied even to consenting adults in private. One would have thought it was drafted and passed through Parliament in order to encourage blackmail: such indeed could have been predicted of it and such, naturally, was its effect.

Apart from the early days, when the Church was dominant, homosexuality has not been a crime in western Continental Europe. The fact that it has until recently been so in England still has repercussions there: one of social ostracism. That attitude, though diminishing, is yet apparent. It is to be combated by knowledge; that the condition, though partly affected by the environment, results from normal variation in the ordinary processes of sex determination. But perhaps the average of the populace will always resent the exceptional, whether the homosexual or the genius.

Sex Linkage

One aspect of sex determination is due to the unusual distribution of the genes carried in the X and Y chromosomes. These are said to be 'sex-linked' (Figs. on pp. 39, 45–6), since they are distributed relative to sex, which those in the autosomes are not. Sex linkage is of three kinds, and to explain this a few words must be said about the structure of the sex chromosomes.

There is a part of X, usually much the larger, which is unique to it in the sense that it cannot be transferred to Y by crossing-over; but crossing-over can occur in that same segment between two X chromosomes when both are present. Secondly, there is the unique section of Y in which no crossing-over whatever takes place. It carries the decisively male-determining genes in man. Thirdly, there is the region, often very short, which is identical in X and Y, so that in it a chiasma holds the sex chromosomes together as a pair in gamete formation. They can then segregate, one passing into each sperm or egg. It will be realized that this pairing portion of X and Y is effectively autosomal, so that it cannot take part in the determination of sex.

When sex linkage is referred to without further distinction, it always relates to the region unique to X, within which crossing-over can occur between XX but not between XY. The following diagrams show how these are distributed. We can now illustrate the matter with reference to colour blindness; a topic of real interest in itself, which can then be discussed.

Though this is an oversimplification, we can first of all think of colour blindness as due to a gene *c* (its allele, allowing normal colour vision, being denoted *C*); *c* is recessive in effect, which means that the action of one *c* is obscured by one *C*. Therefore *Cc* women have normal colour vision, while the *cc* type are colour-blind. Evidently *c* is always effective in the XY sex, in which there is no second X where *C* can be carried: dominance and recessiveness have no meaning in that particular state.

The genetics of colour blindness can be demonstrated as follows. A colour-blind man may marry a normal (homozygous) woman:

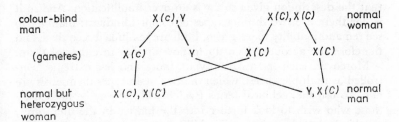

All the children of this marriage perceive colour in the ordinary way; but all the daughters, though none of the sons, are heterozygotes for colour blindness. What will happen when such heterozygous daughters marry a man with normal vision? Evidently a 3 : 1 ratio will result from this union, but one relative to sex, in that half the sons but none of the daughters will be colour-blind.

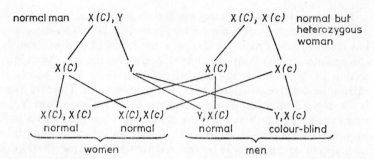

But what is the result when a colour-blind man marries a woman with normal sight but heterozygous for colour blindness? An illustrative diagram would here be almost an insult. For anyone can construct one along the lines just used, and obtain the answer that the four types will appear in equality among the offspring: normal and colour-blind men, normal and colour-blind women. Thus colour blindness can arise in women, though obviously less often than in men; for in women two genes are needed to produce the effect, but in men only one.

We therefore reach the question: how much commoner in men than in women is colour blindness, or any other sex-linked recessive? In fact the two are related as $x : x^2$. So that if one man in 12 be colour-blind, one woman in 144 will be so. On the other hand, a *dominant* sex-linked condition must be approximately twice as common among women as among men: dark-brown eye colour is an example.

Now before returning to the genetics of sex-linkage, it is necessary to examine colour blindness a little more closely. It has been said that the description given so far is an oversimplification. And so it is, for there are two distinct types of colour blindness: inability to see red and inability to see green. Both are sex-linked recessives, and therefore the account given up to now applies to either of them.

Moreover, each has two subtypes, more and less extreme, controlled as multiple alleles, which the two main types themselves are not. To use the technical terms (which will anyway be needed by those who wish to look further into the matter in the literature), mankind can have *protanopia* (inability to see red) or *protanomaly* (the same but less severe).* There is also *deuteronopia* (inability to see green) and *deuteranomaly*. Curiously enough, the protanoid and deu-

---

* Those with protanopia cannot see red, but can see green and blue. Those with protanomaly can see all three, but require an excess of red light to combine with the other two to give the effect of white. The situation is comparable in the two deuteranoids.

teranoid loci on the X-chromosome have not been built into a supergene, or have perhaps diverged from that condition. They are about 9 cross-over units apart, and one human locus, with an entirely different effect, is known between them.

Of the four types of colour blindness, deuteranomaly is the commonest, accounting for about 60 per cent of the whole. Throwing all four together, about 8 per cent of men are colour-blind in western Europe.

The Ishihara tests for colour blindness show that those affected by the condition can actually detect certain colour differences imperceptible to normal people. It is indeed instructive to go into the countryside with an accomplished but colour-blind naturalist and notice the almost uncanny power of seeing cryptic species which he will occasionally display. My own attention was drawn to this by a friend who could readily pick out the fully grown caterpillars of the Emperor moth, *Saturnia pavonia*, six metres or more away on a moorland. Yet though the size of a man's little finger, they are so perfectly adapted in colouring to their food plant, heather, that in general they can be distinguished only by close and careful inspection.

Birds can see colours; but man and the great apes are unusual in being able to do so, for it seems that most mammals are totally colour-blind. Total colour blindness does indeed occur as a great rarity in man. Genetically it is quite distinct from the normal red and green forms, though also recessive.

Numerous sex-linked conditions are known, and in a wide range or organisms. In some the variant forms are dominant, but the greater number are recessive, inherited as is colour blindness. In man, one of these latter is a blood group (pp. 139–48), while another has become so famous that it deserves special mention. This is haemophilia, in which the clotting of the blood is delayed. Ordinary cuts are not dangerous, nor are operations if the patient receives a blood transfusion (pp. 141, 144–5) first. But, following a slight injury, blood may ooze for many days from a mucous membrane, as from a gum at the edge of a tooth. Also, bleeding may occur internally, while blood makes its way into the patient's joints, causing stiffness and pain. Haemophiliacs generally die before the age of reproduction, but a few live to be twenty-five or thirty, and have children.

The genetics of haemophilia follow the same lines as those of colour blindness, from which they can therefore be worked out. Thus a woman who carries the gene for it is normal, but half her sons will suffer from the disease. Those men who do not, cannot transmit it; but though all their sisters will be unaffected, half of them will be heterozygotes. It was at one time supposed that the gene is lethal in double dose, so that haemophiliac women do not exist. This is now known to be untrue. Affected women have certainly survived: they constitute half the daughters from a marriage between a haemophiliac man and a normal but heterozygous woman.

The disease has become famous owing to its occurrence in the royal families of Europe. Queen Victoria was a carrier; quite possibly a mutant. Of her four sons, one (Prince Leopold) was a haemophiliac, while the others were normal and therefore unable to transmit the gene. Since these included King Edward VII, the English royal house is now free from the taint; but its occurrence among the descendants of Queen Victoria's daughters has become a matter of history.

Colour blindness is of course linked with haemophilia, both their genes being carried in the X-chromosome. The cross-over value with deuteranopia (green colour blindness) is 12, and 21 with protanopia (red colour blindness).

I have in the past considered the danger to which the female descendants of haemophilia patients are exposed (Ford, 1973, p. 51), in the sense that they may be heterozygotes. It may well be felt that their sisters ought not to have children. The chance that a woman is a carrier is reduced by half for each generation since the most recent occurrence of the disease: one in four if she has a haemophiliac uncle, one in eight for a great-uncle. It seems convenient and fairly safe to consider that she is free from the gene if she has had eight normal brothers and uncles since the disease last appeared in the family.

There is a rarer and milder form of haemophilia due to another gene at the same locus, and several other conditions with very similar symptoms are known. Of these, Christmas disease is also a sex-linked recessive, while von Willebrand's disease is heterozygous and autosomal.

It will be realized that the distribution of sex-linked recessives is reversed in those groups (birds, and butterflies and moths, for example) in which the Y chromosome is carried by the female. Indeed the first instance of sex linkage to be discovered occurred in the Currant moth, *Abraxas grossulariata*, in which a pale form, *dohrnii*, appeared in half the female, but none of the male, offspring of a pairing between normal specimens. Its subsequent inheritance was similar to that of human colour blindness, but reversed.

We can now turn briefly to the two other types of sex linkage. Apart from very rare chromosome abnormalities, genes totally linked to the Y chromosome are passed down the male line of the XY (or heterogametic) sex indefinitely without transference to the XX (or homogametic) one, in which nothing of the kind occurs.

In the human species, the male determinants, being those decisive in sex control, are inherited in this way (p. 40). So too are the genes for egg colour and pattern in the cuckoo, at least in the European species (p. 128). Here, of course, the XY condition is female.

Finally, there are the genes showing 'partial sex linkage' because they are carried in the pairing region of the X and Y chromosomes. The genetics of this state present certain difficulties, but an important example of it will be mentioned on p. 165. However, total colour blindness in man, just mentioned, seems to be of this kind. In studying the few genealogies in which it occurs, one finds it commoner in males than in females in some pedigrees, and the reverse in others. This is the situation which must arise when a sex-linked gene is sometimes transferred between X and Y by crossing-over.

Sex-controlled Inheritance

In sex linkage, as we have seen, certain forms are associated with sex on the mechanical basis that the genes controlling them are carried in the sex chromosomes. A somewhat similar situation is brought about by different means in sex-controlled inheritance; for this is in a sense 'environmental'. That is to say, certain genes produce their effects only in the bodily condition provided by the male or the female. The interaction concerned may be at different levels, due directly to the sex genes or to the sex hormones (p. 42) that the sex genes have evoked.

Now it is evident that the primary and secondary sexual structures* must be determined by genes sex-controlled in their action. So too must the accessory sexual features: that is to say, those which distinguish the two sexes without promoting the act of reproduction itself. Here one would include the functional breasts of women and the hair on the face of men. The latter attribute is suppressed by the female hormones. These are much reduced after the age of reproduction, so that old women often show considerable traces of beard and moustache.

However, certain genes having no direct connection with sex have yet been selected to operate in the environment provided by the male or female body. We find that situation widely in butterflies

* The primary sex organs are the gonads: the testes (in which the sperms are formed) and the ovaries. The secondary sex organs are the structures required to bring about fertilization.

and moths. For instance, the Clouded Yellow butterflies (e.g., *Colias croceus* and *C. philodice*) have one form of male with, on the upper side, a yellow ground colour and black markings; and two forms of females: one has the colouring of the male, while in the other the yellow is replaced by shades ranging from pale primrose to white. The difference is controlled by a single pair of alleles, with the pale phase dominant. These genes are autosomal and transmitted equally by the two sexes, but that for the pale coloration operates only in the female. The arrangement is indeed widespread among animals. A human example is seen in frontal baldness; an autosomal dominant condition almost restricted to men, though occasionally detectable in women homozygous for it.

A great many features also are partially sex-controlled so as to be much commoner in one sex than the other: the well known human disease of osteoarthritis, a heterozygous condition with a heavy female excess. On the other hand, glaucoma, a common eye defect, is more frequent in men than in women (p. 158), though autosomally inherited.

The Sex Ratio

The widespread necessity for two sexes is due to the need for heredity of the Mendelian type, in which great genetic variability and great genetic stability are compatible with one another (pp. 55–7). For this cannot be attained by mutation. Nor can it be by self-fertilization; since when that takes place, all the alleles that are homozygous remain so, while half of those that are heterozygous become homozygous at each generation.

It will therefore have been noticed with surprise that though a mechanism exists by which males and females can be produced in a clear-cut fashion, in general without intermediates, the detailed working of the system is subject to a number of totally unnecessary alternative arrangements. Any one of these, if extended, would serve for the whole. We are here faced by an opportunism unlike any system imposed by a single master mind. We need not be surprised then that evolution is immensely more often a failure than a success.

Now on looking at the next step in sex determination, the sex ratio, we encounter what at first sight appears another example of what we have just seen: a failure to exploit without complication a device simple and sufficient in itself. In mankind, as in many other organisms, the social structure demands something like effective equality in the two sexes. This the XX, XY chromosome mechanism is perfectly fitted to supply. Yet gratuitously enough, a set of imposed adjustments ensures that the required result should not be attained, after which the situation is repaired to provide what could have been done in the first instance.

Consider the human system which, indeed, is the one we are chiefly concerned with in this chapter. In spite of the XY device, ensuring the production of male- and female-determining sperm in equality, there is certainly a decided male excess at conception (the 'primary' sex ratio). The nature of the situation makes it impossible to determine its precise extent. However, males are slightly in excess at birth (105 : 100) and, as abortions and miscarriages as far back as can be sexed show an increasing proportion of males, it looks as if the original value must be at least 125 : 100. Since only about 76 per cent of conceptions give rise to live births even in civilized communities, there is considerable opportunity for adjusting the proportion of the sexes by differential survival during pregnancy. The female foetus survives better than the male, and women survive better than men. Consequently the equality which could have been obtained at the start is reached later, during the period of maximum sexual activity; between sixteen and thirty years of age in western Europe. But the tendency favouring female survival continues, so that there are actually twice as many women as men among those aged eighty-five and over.

Several problems confront us here. We may first ask: why in mankind is the female expectation of life better than the male?

From the environmental point of view, men on the whole lead the more hazardous lives; though against this must be set the dangers of childbearing. But clearly we are dealing with something more basic than these, neither of which can be active before birth nor, indeed, in the elderly. Evidently it is necessary to look further, to physiology and genetics.

Doing so, we realize that sex linkage itself provides a satisfactory explanation of the problem: it is indeed only on closer scrutiny that we see it is an imperfect one. It must be pointed out that the disadvantageous effects of genes tend to become recessive (pp. 99–101) and are therefore seldom manifest, since they must be present in double dose before they can be detected, *except* in the X chromosome when partnered by Y. This is the male in man and, as far as we know, in other mammals. There is here, then, a built-in system by which the harmful effects of genes in one chromosome always affect the male but not the female. The truth of that cannot be in doubt; but ' "What is truth?" said jesting Pilate' and, unlike him, we may wish to stay for an answer. Now that seems easily obtained from the fact that in some organisms (birds and Lepidoptera) the XY sex is the female; yet there then seems no real indication that the male is the longer lived.*

It appears indeed that the effect of disadvantageous recessives has largely been obliterated by selection. This has either modified their expression or eliminated the genes responsible for them in a situation which involves something like half the population.

An additional test of this matter is available; though, strangely enough, it seems never to have been considered. The effect we have in mind is going to be much greater when the sex chromosomes are large and form one out of a few pairs than when they are merely one out of some dozens, and consequently contain relatively few of the genes. But neither does this source provide evidence of the kind we seek. At this level of the problem, the inevitable impact of sex linkage on the sex ratio just is not evident. We must look further.

It is probable indeed that this distinction in longevity is to be traced to the higher basal metabolism of men than of women; that is to say, to their greater expenditure of energy, which adversely affects male survival. For this view, there is experimental evidence in other animals.

We now turn back to another problem already raised. Why does the sex ratio at conception depart from the equality which the chromosome mechanism is so perfectly fitted to produce? The sexual balance is, like any other feature, subject to genetic variation, and it is possible therefore to alter it by selection; indeed to any degree, up to the elimination of one sex or the other if necessary. Such an adjustment may be attained by the differential survival of the X- or Y-bearing sperms, or else by affecting their power to penetrate or to fertilize the egg. In this way, selection can favour the sex ratio at conception to whatever degree is needed to produce equality, or any other required proportion of the sexes, at the period of maximum sexual activity.

Were that result to be reached by the direct chromosome mechanism, it would not be labile: capable of adjustment to social, environmental and psychological differences. This of course is possible, but only by the devious path outlined.

In the sex ratio, then, we find an example of the way in which genetics can meet and deal with difficulties beyond ordinary expectation. Has the matter, we may well ask, effects of general interest in human populations? We shall not be disappointed in a search for them; but first it is necessary just to mention how the value of the sex ratio is normally quoted.

* It has been held that there is some slight sign of this, but it is not significant.

The *primary* and *secondary* sex ratios are those at conception and at birth. The frequencies of the sexes are represented in either of two ways: as the number of males to 100 females or as the percentage of males. Thus a sex ratio of 100, or of 50 per cent, indicates equality. The ratio is spoken of as 'high' or 'low' depending on whether males or females are in excess.

Now since men survive less well than women, anything contributing to good health tends to produce a high sex ratio. It is lower among the children of elderly than of young mothers, for the latter provide the better prenatal environment; which, in fact, deteriorates not only with age but also from the effects of a succession of pregnancies, particularly when following each other closely. Thus the sex ratio is highest for firstborn children. Since abortions and stillbirths are for obvious reasons particularly frequent among illegitimate children, these have a low sex ratio; so do the offspring of those living in crowded or slum conditions. Indeed a move towards a higher value is then a measure of success in social services.

There is a tendency among the ill educated, or even among the ill informed, to seek a miraculous rather than a rational explanation of what seems mysterious. It is fairly well, though not decisively, established that the sex ratio rises during and immediately after a war; as if some superhuman force were attempting to repair the loss of manhood. That, if at all, it does so with extreme inefficiency matters nothing to newspaper reporters seeking copy; nor did it matter to a bishop of the Church of England who in my hearing drew attention to this as a wonder: almost as an example of divine intervention in human affairs.

If indeed a fact, as it may well be, this wartime adjustment between the sexes could follow naturally from an excess of firstborn children, with their normally high sex ratio, owing to the frequent postponement of marriage or of childbearing during hostilities, as well as to the wider spacing of births at that time. Whether imaginary and spectacular, or reasoned, explanations are preferred depends upon the education and mental attitude of those who discuss them.

3. Evolution

The Speed and Mechanism of Evolution

It is generally supposed that organic evolution is exceedingly slow, requiring many thousands of years to produce a detectable effect. This is an error, and there are several reasons for it. In part, it is an unjustified deduction from the fact that the evolution of one major group of plants or animals from another, say that of the mammals from the reptiles, requires millions of years. So too, the evolution of even a new species is generally, though not always (pp. 112–14), extremely slow.

In consequence, the speed of minor evolutionary changes has been vastly underestimated, while the correcting effect of precise information on the matter has been lacking. This is curious. Evolution is the basic concept of biology, yet it is only in recent years that it has been studied by the basic techniques of science, those of observation and experiment. Some work on those lines was indeed carried out even from the 1914–18 war (pp. 59–60), but to a very limited extent. Such an obvious and serious omission is not easy to understand, and it cannot be allowed to pass without brief comment.

Charles Darwin himself appreciated the need of a practical approach to the study of evolution. Indeed in the mid-1920s his son, Major Leonard, told me of a conversation in which his father had expressed the view that, using a carefully chosen species reproducing annually, the actual process of evolution taking place in nature might be detected in so short a time as fifty years. As a result of some work I was already engaged on at that date (pp. 60–1), it was clear to me even then that Darwin had been much too pessimistic.

Now biologists in the generation immediately succeeding him had explored no such possibilities. They were for one thing too intent upon demonstrating the reality of evolution, for which, perhaps, there was still some need; but, far more, in ensuring that classification should correctly reflect relationship. To them, moreover, though they were reluctant to admit it, such an approach was a more congenial one, because it was in the tradition of their own and of the preceding period. Moreover, strange as it now seems, the rediscovery of Mendel's work was by no means thought to offer opportunities for the study of selection or of evolution in general. Thus *Mendelism and Evolution* was considered an almost revolutionary title for a small book of mine when published in 1931. Indeed it was not until the rise of ecological genetics, mainly during the last twenty-five years, that the actual process of evolution in natural conditions has been at all widely subjected to observation and experiment.

A few examples of that work will be described in this and the next two chapters, after one obstacle of a theoretical kind has been cleared away. But even before that is done, I cannot resist mentioning an instance which, like an introductory text, gives an advance indication of what is to come.

In 1930 R. E. Moreau used his extensive knowledge of taxonomy to estimate the time required for the evolution of a new geographical race in birds. Subsequently, it was found that just such a distinct race of the house sparrow, *Passer domesticus,** a species introduced into North America from Europe in the middle of last century, had evolved round Mexico City, which it reached about 1933. The *minimum* time required for such a change had been estimated by Moreau as 5,000 years. When actually observed in the house sparrow, it was found to have taken thirty years.

Attention must be drawn to two related theories of evolution which have been much discussed, because if acceptable, which they are not, they would have had profound consequences for mankind and other organisms. Evidently, therefore, they must be cleared away at the outset of this or any other consideration of the subject. These are first the idea, due to Buffon (1707–88), that the effects of the environment upon the body can so influence heredity that they can be reproduced in subsequent generations in the absence or reduction of the environmental stimulus which originated them. The second is that of Lamarck (1744–1829), who claimed the same for the effects upon the body of the use and disuse of its parts. These two speculations are often uncritically grouped together under the title of 'the inheritance of acquired characters'. Buffon's view seems to have been dear to the heart of educationalists who would like to feel that their work has in some measure a more lasting effect than through transference by the written or spoken word.

Here we need for a moment to recall the first chapter (pp. 16–17). Genetics must ultimately be judged by its impact upon evolution, in which living organisms are confronted by opposed requirements of what might seem an irreconcilable kind. That is to say, they must possess both great heritable *variability*, upon which selection can act,

* To be distinguished by its colouring and the shape of the beak.

and great heritable *stability* to preserve those qualities and combinations of qualities of value to them. Yet, surprisingly enough, that paradox has not proved an insoluble difficulty. It is overcome by the Mendelian mechanism, which requires that the heritable units (genes) should be extremely constant though they can be shuffled about into new combinations by segregation, yet held together in small groups when necessary (pp. 82–3).

It is essential, therefore, that the genes should be very permanent. We have seen one of the devices that secures this: they preserve their identity without contaminating one another even when brought together in the same individual (p. 17). But that alone is not good enough. To have permanence, they must themselves be intrinsically stable: that is to say, they must not be unstable compounds in the chemist's sense. Yet again, how difficult it might seem to attain what is then required; for they must have some slight degree of instability so that the alleles can exist in different phases, such as those for short fur and for long in a mammal. Such instability of the genes themselves and of the chromosomes that carry them represents *mutation*.

Evidently this must be extremely rare, or it would threaten the essential permanence of the hereditary material. Indeed its average occurrence is too low to measure: a fair guess for a reasonable value is that a given gene may mutate in one individual in a million. But we know that its frequency is not the same for all. Indeed it may become what is for that phenomenon quite common: for instance one in 80,000, as for the disease haemophilia in man (pp. 47–8, 156). Even so, an event which occurs in one individual in 80,000 is by all ordinary standards most unusual. Such rarity is a product of evolution, for mutation rate is under genetic control (Ives, 1950; Westerman and Parsons, 1973).*

Mutation cannot bring about evolution unless it gives rise to some advantage, for the spread of an unassailably disadvantageous gene will be opposed by selection. Nor can a gene neutral as to advantage and disadvantage establish itself, for its spread is then exceedingly slow. Indeed the number of *individuals* in any population which possesses it cannot greatly exceed the number of *generations* since its occurrence, if it be derived from a single mutation, and by that time the accurate balance required for neutrality will long ago have been upset.

It is clear, then, that we cannot appeal to mutation as an evolu-

---

* It is true that a few 'unstable genes' are known. They are quite exceptional, being those which lie in or near a small specialized region of a chromosome containing *heterochromatin*. This in some way enhances mutation, for a normal gene moved near to it becomes much more mutable.

tionary agent. But that is precisely what the theories of Buffon and of Lamarck invoke. For they depend upon external agencies to produce changes in the hereditary material: changes of such a kind, moreover, as to reproduce in the body, through all the complications of development, those very effects to which the original environmental stimulus itself gave rise. We can then exclude the 'inheritance of acquired characters' as an evolutionary mechanism. Mutation is ultimately responsible for the diversity of the alleles necessary to produce genetic variation, but the genes must be much too stable for it to *control the course* of evolution (Fisher, reference p. 69).

For that purpose we require *selection*: an idea normally understood today, so widely have Darwin's views become part of the intellectual stock-in-trade of ordinary educated people. However, there are a few aspects of it which need slightly to be clarified.

Natural selection ensures that individuals with an overall advantage tend to contribute more to the next generation than those less well endowed. We say 'overall' not merely because their advantages must outweigh their inevitable disadvantages (and see also p. 79) but because qualities that are an asset at one stage in the life history are sometimes a handicap at another (pp. 63–4). Individuals of course vary among themselves; or there would be nothing to *select*, and survival would be at random. If the variation on which selection operates is genetic, it can be accumulated and spread through the population, causing it to change and evolve.

But everyone knows that all variation is not genetic; normally about half of it is environmental (pp. 31–4). If we grow plants in different soils or breed butterflies at different temperatures, the offspring may not look alike. That effect will persist for as many generations as the environment in question, but not afterwards. Yet this does not mean that the subsequent population will not respond to the novel experience of their ancestors. For they may have become adapted to the peculiar conditions they encountered by selection operating on the genetic as well as the environmental aspect of their variation. Such a result is often a disaster, for what may be an advantage at one temperature, for example, will not necessarily be so at another, and it is difficult to reverse an evolutionary trend when no longer favoured. Evolution, indeed, is vastly more often a failure than a success.

If we keep organisms in strictly constant conditions in the laboratory, all their variation is genetic. If we study a group of individuals that are genetically identical (a 'pure line')*, all their

---

* This can arise in those forms, chiefly plants, which can be self-fertilized. For their genetic variation is then halved at each generation, because the

variation will be environmental. The environment may affect the action of a gene, but not the gene itself.

There may sometimes be confusion between the two types of variation (pp. 31–4). Yellow pigments, known as xanthophylls, generally present together with the green colouring matter in higher plants, are stored by rabbits in their fat, but in a colourless form because they are oxidized by an enzyme produced by the action of a gene Y. Therefore their fat is white. The recessive due to the action of the alleles yy prevents the formation of this enzyme, and consequently the fat becomes yellow in such animals when supplied with green food. If this is not provided in the diet, the two white-fat conditions cannot be distinguished in appearance. Thus in the presence of green food, white and yellow fat are genetic variations in normal rabbits. In the yy type the same two conditions are environmental variations.

There is another matter that must be made clear at this point. Organisms have thousands of genes, and a considerable proportion of their alleles is heterozygous (pp. 107, 153). In view, once again, of the permanence of the genetic material, if all the progeny of a plant or of an animal were to survive, they would show an undue amount of variation (even though many of the features available for study are chemical or physiological, not evident on superficial in-spection: see pp. 72, 82–4). Thus if a population remains constant, or relatively so compared with another, that constancy is the result of selection and is an aspect of evolution. Genetic variation is therefore an equilibrium condition between segregation, and to a minute extent mutation, tending to promote it; and selection, tending to reduce it. That is why, for instance, seedlings are often more variable than mature plants.

Evolution at the Present Time

With these points in mind we can go ahead and examine what Darwin would have given so much to do, the process of evolution taking place today in wild populations. As the need for additional genetic concepts arises, they can be explained, painlessly, as it were, in passing.

The possibility of observing evolution and adaptations and sub-jecting them to experiment is an outcome of *ecological genetics*. I have heard it said that there is no essential difference between

------

offspring of homozygotes remain homozygous for the pair of alleles in question, while half the progeny of the heterozygotes will be homozygous.

that study and population genetics. Yet the difference is a fundamental one. Population genetics principally involves mathematical
analysis, often carried out with little or no regard to organisms in
a state of nature. Such calculations may therefore employ assump-
tions as to population numbers, genetic interactions, adaptations,
conditions of the environment and selection pressures; and these,
being based largely upon inference, have frequently proved profoundly inaccurate.

Studies of that kind, if founded upon precise information, can be
one of the techniques of ecological genetics which consists, so to
speak, of taking genetics out into the countryside and combining
the results of scientific natural history with genetic and other
laboratory experiments. For this, statistical methods are of course
essential. We must calculate the errors involved in our observations,
and whether or not separate samples can be combined to produce
a sufficient total. And it is sometimes necessary to make, perhaps by
the methods of population genetics, theoretically devised models;
not that these are likely to reflect the true condition, but in order
to determine by the techniques of ecological genetics in what ways
they depart from reality. Moreover, it is frequently essential to
decide whether the products of segregation have equal survival
values. It is to ecological genetics, therefore, that we must turn in
this and the next two chapters.

When it was realized that an important but neglected aim in
biology is the study of evolution in progress by means of observation
and experiment, there seemed a need to identify situations in which
that process is exceptionally rapid. For only as a result of such work
did it transpire that natural selection is often very powerful.

Bearing in mind, then, that genetic variation is in equilibrium
between segregation on the one hand and selection on the other,
special attention was given to populations that fluctuate greatly in
numbers. The first instance of the kind to be examined (by my
father and myself, Ford and Ford, 1930) occurred in a butterfly,
the Marsh Fritillary, *Melitaea aurinia*.* This has one generation
annually, and flies in late May and most of June. It feeds on the
Devil's-bit scabious, *Scabiosa succissa*, and forms isolated colonies
mainly in marshes. One of these occupied a few acres in Cumber-
land, England, where entomologists had captured large samples
from 1881 onwards. Their collections and notes had been preserved,
and consequently were available for study in after years.

* There are several advantages in using the scientific names. Only one
need be mentioned here; they are international. To what English butterflies
does a Frenchman refer when he speaks of 'Robert le diable' or of 'le grand
Mars'?

Up to 1897 the insect was immensely abundant there, flying, as
the records say, 'in clouds', and was remarkably constant in appearance each season, year after year. Subsequently the numbers began
to decline, and by 1912 the species was quite rare in the locality.
Then, for the next seven years, very few were to be seen. A search
of some hours at the height of the emergence would produce only
half a dozen specimens or so; and they were of the same form as
before. Yet in 1920 and the four years following, a great numerical
increase took place; and indeed, as a single female lays about 200
eggs, the opportunity for this is evident. In fact, by 1924 there was
a dancing haze of the butterflies in the few marshy fields to which
they were restricted. At the same time they underwent an outburst
of variation. Not only were they extremely diverse, but many of the
more exceptional forms were clearly at a disadvantage: their wings
tended to be crumpled or asymmetrical, and some could hardly fly.
From 1925 to 1936, when the observations ceased, the population
stabilized, and indeed the butterflies gradually became less common. At the same time they were relatively invariable once more,
and only one or two of the highly unusual specimens were seen.
Consequently the Marsh Fritillary had in that locality returned to
a constant and indeed a permanent form, but this was not the one
which characterized the colony before its outburst of abundance.
That is to say, the insect had passed through a period of numerical increase with high variability, as a result of which it had
evolved.

The more striking and distinct varieties probably segregate on a
simple basis; indeed two of them are known to do so, as recessives.
However, most of the observed variability was 'multifactorial': due,
that is to say, to a number of genes having small, similar and
cumulative effects; as with the genetic aspect of human height.

An occurrence of this kind has several times been reported subsequently in this and in other species. What are we to think of it?

When numbers increase, some aspect of selection has been relaxed.
This allows the spread of forms which would not have survived in
more rigorous conditions, so that the population becomes more
variable. The reverse is true when stricter selection sets in again.
At that time, there is heavier elimination, producing both smaller
numbers and a more standardized type; for only the best adapted
individuals survive. Consequently, numerical expansion with diversity prepares the way for numerical decline with uniformity. But
why should this sequence promote evolution? The reason needs
brief consideration in terms of genetics.

Genes have multiple effects. It is not true that there is a gene
merely producing, for instance, brown colouring with an allele

producing white. The gene influencing colour will act also in distinct and diverse ways: on, say, pattern and length of life. Well over 1,000 genes must by now have been studied in the fly *Drosophila melanogaster* (p. 28); and many, though by no means all, have been identified in quite trivial ways: by a slight change in eye colour or wing neuration, for instance. Yet there has never been one of them which does not have an important effect on the working of the body: affecting, perhaps, the activity of the insect, the number of eggs laid or ability to withstand unfavourable conditions. Even so, a trivial and easily observed feature may often be used more conveniently to mark the presence of a gene than one of a more important though more subtle kind (pp. 71–2).

In addition, genes tend to interact with one another. We shall encounter this in discussing human polymorphism (pp. 148–55), and we are constantly faced with it in other animals and in plants. For instance, in the fly *Drosophila melanogaster*, brown eye colour is a simple recessive compared with the normal dominant Red; so too is scarlet eye colour. But when these two recessives are brought together, they give rise to white eyes. In view of such interactions, as well as the multiplicity of their effects, the genes must combine to form, as it were, a genetic background, which long ago (1931) I named the *gene-complex*.

Thus, when in *Melitaea aurinia* relaxed selection allowed less favourable genes to spread, there was an opportunity for them to be tried out in a great variety of combinations with a few of which they might interact in a beneficial way: a result that would take an immense time to achieve in a numerically constant population.

That instance was one in which a colony underwent a long-term fluctuation in numbers; but many organisms are subject to extreme yet regular numerical changes which must provide also opportunities for rapid evolution. Thus the fly *Drosophila pseudoobscura* in California is annually reduced to very small numbers in the winter. It has about seven generations through the spring and summer, during the first two or three of which it becomes vastly abundant. This certainly gives an opportunity for evolutionary adjustment. Of that, an instance is provided by a chromosome variant known as CH (pp. 83–4), which can spread only in the relatively favourable situation of an expanding population.

Another and promising example of a slightly different type is being examined at the present time in a day-flying moth, the Scarlet Tiger, *Panaxia dominula*. This has one generation annually. It occurs in isolated colonies in southern England and Continental Europe, sometimes very commonly. The population of this species in a marsh at Cothill, five miles south-east of Oxford, has been studied as to numbers and other features of the perfect insects, every season for thirty-seven years.

The numerical aspect of that work is carried out by the technique of marking, release and recapture which, employed daily over the three- to four-week period of flight, involves mathematical treatment; for it necessitates calculating the average length of life during that time. However, the idea behind the method is very simple. If we catch a considerable number of specimens, mark them with a waterproof paint, release them and allow them to mix thoroughly with the colony, an estimate of the total population-size can be obtained from the proportion of marked individuals found in a second sample. Thus if 100 insects are so marked and liberated, then if among 80 caught subsequently 10 are found to be marked, we calculate the population as 100 × 80 ÷ 10 = 800. The work is continued over successive days and we have, in reality, to know the date on which each specimen was marked, or received multiple marks. That information is easily supplied by varying from day to day the position and colour of the dots of paint (the purpose of this study is further explained on p. 70).

In combination with that technique, the number of caterpillars on the site earlier in the year can also be established, but far less easily. A sample of them is marked and redistributed proportionately as found. It is then necessary to determine the number of moths, taken on the wing, to which these have given rise. One cannot of course by ordinary means mark a caterpillar so as to identify in nature the perfect insect it produces. This, however, can be done by making the caterpillars radioactive* and scoring the moths some weeks or months later by means of a Geiger counter. That work has already been carried out successfully by Kettlewell and by Cook. They both found that approximately 90 per cent of the larvae die on approaching pupation. That remarkable result accords well with the vast numbers of caterpillars, compared with perfect insects, in colonies of this species, the females of which lay 200 to 250 eggs. Here we have a constantly recurring opportunity for evolutionary adjustment on the same lines as that which took a period of years in the butterfly *Melitaea aurinia*. Has it, in fact, the same effect? We may confidently expect so. We hope to be able to answer that question decisively by the use of a method devised some twenty years ago, but not widely employed until rather recently.

Larvae of the Lepidoptera (butterflies and moths) seem to have but few obviously variable features, but some aspects of their

---

* Sulphur-35, emitting low-energy β rays, with a half life of 87·1 days, proved to be a suitable isotope for this purpose.

chemical diversity can now be examined. This can be done by means of *electrophoresis*. Mutation produces chemical changes in the protein which a gene controls. In a proportion of instances this affects the electric charge of the substance in question, causing it to migrate faster or slower in an electric field. Such migration can be made to take place in a starch jelly or other somewhat similar medium, and can then be made visible by suitable stains. This technique is already suggesting much greater variability in the caterpillars of *Panaxia dominula* than in the flying moths. These latter comprise the more favourable variants that have survived from the earlier stages.

It will be evident that when a species is introduced into a new region, or arrives there owing to an exceptional act of migration, there will be great opportunities for its numerical increase, should it establish itself successfully. Moreover, it will be faced with an unaccustomed environment to which it must become adapted or perish. There is here an opportunity for observing and studying rapid evolution. Of this an example has already been mentioned (p. 55) in regard to the European sparrow in the north-eastern United States, where it was introduced from England and Germany in 1852.

Evolution in the Meadow Brown Butterfly

We can now turn to a different but fruitful field for the study of evolution and of natural selection: that is to say, the situation in which a plant or animal has adapted itself to highly diverse or changing conditions by means, especially, of multifactorial inheritance (pp. 60, 73). These requirements are met by the Meadow Brown butterfly, *Maniola jurtina*, in which they have been studied in considerable detail. In that species the underside of each hind wing may be spotless or may carry from one to five spots, a variation due in part to a considerable number of genes acting cumulatively. Though present equally in the two sexes, their effect upon spotting is greater in the females than in the males. It increases also with temperature (p. 31).

The Meadow Brown has one generation in the year. At a low altitude, up to 200 m in Italy, it emerges from the chrysalis in late May or early June. Pairing takes place at once, and the males die after an adult life of only five days or so. The females, however, aestivate, remaining quiet in bushes, and so survive the hot and dry conditions of an Italian summer. Early in September they become active again, and are then quite common; though inspection shows

c

that the population is composed almost entirely of females. These had received sperm when they paired three months previously. With this they now fertilize their eggs and lay them on grasses, which constitute the food of the caterpillars.

On emergence from the chrysalis and in early aestivation, about 24 per cent of the females are spotless. When they fly again in September, non-spotted specimens amount to 48 per cent. That is to say, the spotted individuals are subject to a heavy differential elimination while they aestivate. If we consider the elimination acting against those with two or more spots, it amounts to 64 per cent during that time.

Why then do not the butterflies, in that locality at least, become spotless? The reason is that though the genes for the higher spotted types are at a disadvantage in the adults, they are at an advantage in the early stages. Here we have *endocyclic* selection, in which the effect of a gene (or a group of genes) changes from good to bad, or the reverse, during the life of each individual. In *cyclic* selection, on the other hand, that is not so; for the change from advantage to disadvantage, or the reverse, takes place from one generation to another: as in the CH chromosome type of *Drosophila pseudoobscura* already mentioned (p. 61).

It is important to notice that, even at sea level on the Italian mainland, a few females (about 5 per cent) do not emerge until September: and a few males, sufficient to fertilize them, are similarly retarded. In that respect, these approach the situation found in the cooler and moister conditions above 700 m, where even the Italian emergence is in both sexes spread out over a period of weeks, from mid-June to early September, and aestivation does not occur. At intermediate heights, 350 to 500 m, about half the population adjusts one way, by aestivation; and half the other way, by means of a lengthy period of hatching from the chrysalis. The situation in England, and north of the Alps generally, is similar to that at the higher altitudes in the Apennines.

A further aspect of this matter is noteworthy. Island habitats tend to be more extreme, partly because less diversified, than mainland ones. The butterfly caters for this in a remarkable way. On the islands off the Tuscan coast (Elba, Giglio), one finds the same aestivating habit as that already mentioned, but carried to the extreme, for it affects all the females: the 5 per cent or so of the two sexes emerging in September are absent. In England, where the butterflies do not aestivate and the emergence is prolonged, as on mountains in Italy, this behaviour is carried further in the Isles of Scilly. There the butterflies continue to appear into September. My colleague W. H. Dowdeswell and I have on numerous occasions left Cornwall in early August to work in Scilly. In both places the Meadow Brown was flying in numbers at that time. So it was in the Isles well into September when we returned, but all had by then disappeared on the mainland. Thus the butterfly adjusts to an island life both in Italy and in England, but by different means.

It was in Scilly that *Maniola jurtina* was first used in ecological genetics; and we have gained extensive experience of it in that exceptional habitat by repeatedly camping there, especially on the uninhabited island of Tean. Indeed, as my colleagues and I have on numerous occasions had to camp in order to carry out scientific studies, it may not be inappropriate now to make a few comments on that way of life for that particular purpose.

In the first place, camping is of two main kinds. There is the temporary camp put up for a night or so to be on the spot where suitable accommodation is lacking or distant. That calls for no special remarks. On the other hand, there is the semi-permanent camp in which one settles down, maybe for some weeks of work, perhaps in isolated conditions on an island, as we have so often had to do.

Such a camp must become a home, and the better run and looked after it is, the better will be the results that come from it. Anyone who knows what he is about camps for scientific work with a reasonable maximum of comfort. Use sleeping bags upon light camp beds. Employ ridge, not bell, tents, with a trench dug around them unless the soil be very sandy, as often in Scilly. Two people require two tents: one for cooking, food storage, meals and ordinary daily use. There should be a box-table and a chair side by side for each person just within the doorway, the cooking arrangements in the centre and stores in strong cases disposed at the sides and back.

The other is the sleeping tent with its two beds. There should be a box-table at the back between them, and it is well if they are high enough above the ground for a suitcase under each. This tent may also be a laboratory where the cleaner experimental work can be done (e.g. marking butterflies). There should be fly-sheets at the ends of the tents as well as along the sides, to avoid 'wet-pillow nights' when rain in a fine spray is driven through the canvas in a gale.

We belonged to the Primus-stove age; perhaps this type is still best in remote conditions. There should be a small oven, with a shelf inside, to fit on it. Enjoyable food and good cooking are most important. Conditions nearly intolerable can be gaily supported if one can look forward to one's meals. A benison goes up to him who returning after a hard day, perhaps on a neighbouring island, prepares with quick efficiency hot scones for tea. The camp must not be polluted by the presence of a drunkard or a total abstainer.

Think not to live off the land: occasional augmentation, yes, but nothing more. After breakfast; clearing up the camp; preparing a picnic lunch; scientific work, perhaps in various habitats, for hours; cooking; a quiet drink before dinner; washing up; probably analysis of results in the evening; and planning for the future, there will be no time for food gathering or fishing. As to fresh meat and vegetables, *solvitur ambulando*. Remember disinfectant tablets for the drinking water; simple remedies; sunglasses; rat traps; and this great rule: semi-permanent camping begins with an unstinting visit to a really good grocer.

Our work on *Maniola jurtina* in Scilly throws light upon evolutionary adaptation, and a few aspects of it may briefly be mentioned here. An account of it will be the more easily intelligible if restricted to female spotting.

Those of the islands in Scilly where we have principally studied the butterfly fall into two size groups: three 'large' of 275 ha or more, and five 'small' of 16 ha or less; a difference of at least seventeen times.* The spotting of *M. jurtina* differs characteristically in these two types of habitat. It has remained fairly constant on each of the three large islands, being similar on all, with approximately equal numbers at 0, 1 and 2 spots: a 'flat-topped' distribution, in which the higher values become progressively rarer. On the other hand, spotting differs widely from one to another of the small islands, though remaining relatively constant year after year on each; except when it responded to a severe ecological change on two of them (pp. 67–8). On none does it permanently resemble the large-island type; for it may have a single maximum at no spots or at two; or two maxima, the greater at no spots and the lesser at two, or the reverse.

The similarity of spot distributions on the large islands, and their differences on the five small ones, has given rise to considerable speculation, often unfounded and made in ignorance of the relevant facts. For it is noteworthy that whenever attention is drawn to instances demonstrating natural selection, attempts are made to explain them on non-selective grounds, and to this the present situation is no exception.

In the first place, we can eliminate one obvious possibility; for we find that the similar spotting of *M. jurtina* on the three large islands cannot be due to the migration of the butterflies between them. It must be noticed that this species is far from a wandering one: save

---

* The large are St. Mary's, Tresco, St. Martin's; the small are Tean, White Island, St. Helen's, Great Arthur, Great Ganilly.

for very rare exceptions, even a hundred metres of highly unsuitable territory proves a complete barrier to it, and there is at least a kilometre and a half of sea between the islands in question. Moreover, in two of the channels between the large islands there are small islands on which one finds butterflies with markedly different spotting types. It may also be noticed that some small islands inhabited by colonies whose spotting is highly distinct from one another are much closer together than are the large ones on which the populations are similar. These facts, on the whole, serve to dispose of migration as a cause of resemblance; quite apart from the one shortly to be mentioned which excludes it.

It has been supposed that the special features of the Meadow Brown in Scilly could be attributed to numerical fluctuations in the past, on the assumption that these can occur more easily in small than in considerable communities, which may well be true. It must be noticed, however, that the rapid evolutionary adaptations to which such fluctuations may give rise are associated with numerical changes, not with total population size.

There is another aspect of the matter to which C. H. Waddington has drawn special attention, though its existence has in fact often been considered. He suggests that on the small islands the insect may in the past have been reduced to a few individuals, or actually have become extinct. In those circumstances, he holds that the gene frequencies of the survivors, or of occasional rare immigrants, could then affect, by chance survival, the local races derived from them. This might produce distinct types on each small island if the genes concerned were of neutral survival value which, as abundant evidence shows, they are not. Waddington suggested that the large islands might retain considerable numbers of the butterfly even when the race was reduced to a minimum, and therefore perpetuate on them an original Scillonian type; so resulting in uniformity between the populations inhabiting each of them.

Others have put forward variants of these suggestions; but they are not worth considering, in view of certain facts which dispose of them. It will suffice to mention one of these.

In 1957 Scilly, as the rest of England, experienced abnormal weather. A severe drought started in February and continued until early July. On the sandy soil of the islands, this stunted the bracken and brambles and produced a poor crop of grass while the larvae of the butterfly were feeding. When the perfect insects emerged, it was found that on two of the large islands (St. Martin's and Tresco) a change in spot-frequency had actually occurred; the insects had departed from their characteristic type and become relatively spotless, with a single maximum at no spots.

The vegetation had largely recovered by 1959, and spotting had also returned to normal, with approximate equality at 0, 1 and 2 spots. Had it been known that the large-island spot distribution can shift to another type, it seems unlikely that suggestions would have been made to interpret it in terms of survival from a remote past. Moreover, the fact that the new values did not persist when the exceptional climatic conditions ended, but returned to what they had been before, shows that selection favours those genes for spotting which are best adjusted to the normal environment.

In view of this and of other evidence, we now have an explanation of the differences of *Maniola jurtina* on the two types of islands in Scilly. Selection can adjust the butterflies to the distinct local environment of each small island, while it can do so only to the average conditions on the large and much more diversified ones; and averages tend to be alike. This also interprets the fact that St. Mary's was the only large island on which the spotting did not alter in 1957. For it is the largest and most ecologically varied of them, on which areas of normal vegetation are to be expected even in an unusual season.

It will be noticed that work on the Meadow Brown butterfly shows that this insect is subject to powerful selection, whether preserving its spot frequencies in a permanent form or as a recurrent sequence of advantage and disadvantage within each individual. It might have been objected that this does not demonstrate evolution, had it not been pointed out already that it is only by means of selection that genetic stability can be maintained.

No informed person has ventured to suggest that the distinctive spotting types characteristic of the several small-island populations in Scilly have arisen merely by the fortuitous survival of genes with selectively neutral effects, that is to say by 'random genetic drift' (p. 56), unless they were temporarily reduced to very small numbers, a few dozen or so. In those circumstances such chance effects can be operative, though obliterated by selection as the numbers rise again (see my discussion of the subject: Ford, 1975). And here we know the normal population sizes, mostly comprising many thousands. They have been calculated by the method of marking, release and recapture (p. 62). Indeed the *stability* of the spot frequencies year after year on each small island entirely precludes such an explanation.

However, we thought it right to provide an experimental test of this matter by establishing an introduced population of known size and spot distribution on a suitable small island from which the species was absent. This we did by transporting females (probably almost all fertile, as is usual in the wild) from St. Martin's to Great

Innisvowles, which is small, less than half a hectare being available for the butterfly. The result was not fully conclusive because our artificial colony died out rather too soon to give us a clear answer. But the fact that spotting changed year after year in the same direction, to a higher value, would if continued a little further have excluded change due to chance survival of selectively neutral genes as the agent concerned. A further piece of work of a similar kind is now in hand.

An interesting point arises here, of general significance, whether in science, archaeology, art or general enquiry: the difficulty of establishing a negative. It is clear enough that for the purpose we had in view it was necessary to pick upon an island where the butterfly could survive, having the grasses necessary for its caterpillars to feed upon, but was in fact absent. That is easily said, but many days of what might seem unrewarding work were necessary to demonstrate it. It is ever the same: only laborious efforts or great knowledge will suffice to establish a negative.

A friend of mine, Montagu James, Provost of Eton, was once asked where some simple quotation was to be found in the novels of Dickens, on which he was a great authority. Walking over to the window, he stood in silence for a few moments and returned with his verdict: 'It is not in Dickens.' No one else in the world could have said that with certitude. I doubt if our conclusion on the absence of the Meadow Brown was founded as securely as that of Dr. James on the Dickens novels. Certainly if the insect did occur at all on Great Innisvowles it must have been exceedingly rare and, since its spotting is multifactorial, the contribution made by one or two specimens to the 223 we imported must have been negligible in its effect upon the few generations we were able to study.

As a result of work on ecological genetics it has been found that populations are subject to balanced selective forces very much more powerful than had been predicted on general grounds. When the distinguished mathematician R. A. Fisher wrote his great book *The Genetical Theory of Natural Selection*, 1930, he envisaged selection for advantageous qualities of about 1 per cent in wild populations, and he was criticized for putting the value so high. We now know that it is usual for it to reach 20 or 30 per cent, often very much more. This has an important effect quite apart from its efficiency in maintaining optimum stability. That is to say, it enables a population to adjust rapidly to changing conditions. Of this the immediate reaction of the Meadow Brown butterfly on Tresco and St. Martin's to the unusual weather of 1957, and its subsequent rapid return to normality, provides an illustration. However, our studies of that insect in Scilly have on several occasions enabled us to examine its response to outstanding changes in its habitat. One of these, the details of which were rather complex, can be indicated only briefly here while another can usefully be discussed rather more fully.

For many years my colleague W. H. Dowdeswell and I camped on the uninhabited island of Tean when the Meadow Brown was on the wing. This is one of the larger of the 'small' islands of Scilly, being 14 ha in extent. During 1953–4, the female spotting of that butterfly changed, on one isolated part of the habitat, from having two maxima, at no spots and two, to a condition with a single maximum at the latter value. It did so in response to a profound alteration in the ecology of the area due to the removal of a small herd of cattle long maintained there. The spot frequency of the insect remained constant year after year both before and after that event. The selection required to produce the observed effect could be calculated, using the mark, release technique (p. 62). It amounted to 64 per cent against non-spotted females.

Another occurrence of this kind which we witnessed was on White Island. This has an area of about 16 ha and is bent at right angles into two parts, differing in shape but not greatly in size. The southern, and slightly smaller part is long and narrow with a ridge of high ground, rising to 40 m, running along its east coast. The northern is roughly triangular, with a central hill about 50 m high. The configuration of the two, and especially their alignment in relation to prevailing winds, ensure that they have distinct microclimates. Their vegetation consists of rough grass, with considerable areas of bracken, bramble and gorse. Where they meet, the ground descends almost to sea level, forming an isthmus which separates the higher ground by about 50 m.

We began to collect the butterfly on White Island in 1953, and we continued to do so subsequently. The distinctions, climatic and other, between the two parts induced us to examine the populations on each separately, but their spotting did not differ: the females had a single maximum at no spots throughout the whole island; a condition maintained for the next four years.

However, during a great storm in the winter of 1957–8, the sea washed across the isthmus on White Island, covering it with shingle and destroying the plants that grew there, so forming a barrier which divided the butterfly colony into two. The next summer, that of 1958, we found that these now distinct populations had come to differ. That on the northern area remained as it was, while the one on the southern had changed: the individuals with 0, 1 and 2 spots had become approximately equal in number; a result requiring selection of 58 per cent in favour of the spotted forms that year compared with the previous one.

The situation so produced continued until 1970, when the Meadow Browns on the southern part of the island began to acquire their former spot-frequencies, approximating therefore to those of the northern population. This must surely have been related to the fact that by then the barrier created by the storm of twelve and a half years before was being obliterated by the growth of plants across it, so that the butterfly communities were reuniting.

One cannot refrain from commenting on the circumstances which enabled us to investigate this event so fully. If a biologist were to hear that an island had effectively been cut into two by the sea, he would naturally wish to analyse the result of such a cataclysm. But it might be thought a piece of quite extraordinary good luck that this should have occurred where details of the variation and adjustment of a suitable species had actually been obtained in advance of the event. In fact, this was no mere chance, but a dividend which accrued from our own policy. We were annually accumulating information on the ecological genetics of the Meadow Brown on a number of islands in Scilly partly because they seemed obviously liable to very considerable environmental changes, the effects of which it would be valuable to study in the future with reference to the situation existing in the past.

Having now drawn attention to a few instances out of many in which the butterfly *Maniola jurtina* has been used to detect and study evolution in action in Scilly, it is appropriate to relate it to certain other conditions affecting this species in southern England.

In the first place, one can hardly fail to be impressed by the great power of selection which can operate on this butterfly in natural conditions and, consequently, how quickly the insect can adjust to changes in its environment. It will have been noticed that as so far explained, the number of spots on the underside of the hind wings has been used as the sole criterion of its variation. This is quite a trivial feature: the spots are small, mere dots in fact, and it is likely that their presence is in itself of no significance. Not so the genes controlling them, for we know that these must have important effects (p. 61). Indeed they influence the length of the whole life cycle and, as already mentioned, the survival of the imagines* (p. 64). Further, and surprisingly enough, they affect the frequency with which the caterpillars are killed by certain internal parasites.† These, perhaps, are susceptible to changes in

* The imago is the perfect insect, usually winged, which emerges from the chrysalis.

† The species which attack the Meadow Brown caterpillars and inject their eggs into them belong to the Hymenoptera (the great group that includes bees, wasps and ants).

the body fluids and tissues due to the action of the genes for imaginal spot number. Thus spotting in the Meadow Brown is the outward and visible sign of inward physiological properties. Indeed when other features are studied in addition, its significance is fully established. This was apparent when we examined female spotting across southern England from east to west. The *position* of the spots is subject to a cline,* in which they become progressively more anterior on the hind wings from the east coast to Land's End and thence to the Isles of Scilly. It is reversed, however, over a region from mid-Dorset to east Cornwall. It is also in that same district that a change in *spot number* occurs. This alters from the type found in most of southern England, and indeed thence to the north of Britain, with a single maximum at no spots. It has, however, two maxima, at no spots and two, spots, across Cornwall. The position where that change occurs varies within the range in question and is generally, but not always, abrupt. But a different type of assessment is also related to that same tract of country. It is one affecting two independent forms of chemical diversity in the body, detected by means of electrophoresis (p. 63). This also takes place from mid-Dorset to east Cornwall. The variability which it demonstrates is of a different type from any so far discussed, and introduces a situation to be described in the next chapter (p. 84).

Why should these various features all alter in the same part of England and, on the whole, suddenly in respect of spotting, while moreover the chemical diversity is unstable in the same district but is not known to be so elsewhere? This matter has been excellently discussed by Handford (1973). He points out that, as is well known, many aspects of the environment change continuously and independently along a geographical transect, but it is unlikely that the reactions of plants and animals to them will also be made up of independent items. Indeed they will certainly involve sets of co-adapted genes which cannot act independently of one another. Consequently, the adjustments of an organism to the various requirements of its habitat will tend to be relatively few in number, and distinct. Thus one reaction may displace another more or less abruptly, even when responding to a continuous change. The positions of these genetic discontinuities may alter from season to season, since the genes in question will act in adjusted groups to the diverse ecological conditions; and for that reason the distinctions between them will tend to hold together when they shift.

That type of adjustment has also been reported in the plant

* A cline is a continuous change in some form of variation across an area of country.

Potentilla glandulosa growing from sea level to alpine conditions in the Sierra Nevada of California. Within that range a series of forms exists, adjusted discontinuously to what has been called a continuous change of climate. There is, however, no real evidence here that the change in *environment* is a continuous one. It may well include discontinuities in soil content, and it will certainly do so in respect of the other plant species with which *P. glandulosa* has to compete at different altitudes. Consequently, this instance is a much less convincing one than that provided by the butterfly, in which the clear-cut distinctions can shift from place to place. Yet it is an important discovery that these adjustments in the plant involve as many as sixty to a hundred genes.

Indeed it is clear that selection acting upon multifactorial variation can be rapidly efficacious in adapting organisms to different aspects of the terrain that they occupy. We shall often encounter the fact that it does so more rapidly than when a desirable character must be spread and improved, though due to the action of a single major gene. For this is liable to disturb an already adjusted gene complex, which the gradual build-up of a distinct genetic background multifactorially is much less prone to do.

A telling example of such adaptation is provided by the sea plantain, *Plantago maritima*, which can colonize extremely dissimilar habitats, such as crevices in dry rocks or waterlogged mud in salt marshes, with a series intermediate between them. These entail considerable anatomical and great physiological adjustments, achieved by selection acting upon multifactorial variation. Yet we have here a somewhat unusual form of adaptation. Plants from any of the extreme conditions grow perfectly well in an ordinary garden provided they are freed from competition. It is this which they escape by their ability to maintain themselves successfully in places of an exceptional type. We find the edelweiss, *Leontopodium*, driven to a similar device. It will grow at sea level, but it secures itself from competition, which it cannot sustain, at the edge of the alpine snowline where other species are excluded.

The Origin of Species

In this chapter we are discussing the effects of selection operating on the simultaneous action of many genes, whether controlling features of a discontinuous type or those having many relatively small, perhaps similar, and additive effects. Thus it leads us to consider the origin of species: how do these arise from one another? There can be no doubt that they do so most frequently from an accumulation of gene and chromosome differences which pass from ordinary variation through the evolution of local races or subspecies on to speciation itself.

It has been pointed out (pp. 25–6) that an essential step in the formation of the reproductive cells consists in an attraction bringing together, and during a critical stage in the process holding together, the pairs of homologous chromosomes whose members are derived respectively from the two parents. Also that this attraction is not a property of the chromosomes as a whole, but of the alleles they carry, which are situated along them at definite loci; normally therefore opposite one another.

It seems that this allelic attraction must be stronger when the pairs are identical (homozygous) than when they are dissimilar (heterozygous). Evidently the process of building up distinct adaptations will involve heterozygosity, and ultimately a weakening of chromosome pairing, between the alleles concerned when races evolved to meet distinct local conditions are crossed. These are most easily produced in geographical isolation from one another, as Darwin himself realized.

Moreover, it will later be mentioned (p. 83) that chromosome reconstructions such as inversions, which differently in distinct habitats, tend to provide opportunities for local adaptations because they supply some degree of *genetic* isolation. They do so because they check the pairing of homologous chromosomes by reason of the alleles responsible for the attraction between them no longer being opposite each other (pp. 82–3).

In what way do distinct communities show that they are moving in a direction that leads them to become separate species? As pointed out by J. B. S. Haldane in 1922, any tendency for two forms to speciate will show itself first in the heterogametic sex: that with the XY chromosome pair (e.g. the male in mammals and the female in birds). That conclusion he formulated in his well known law, which states: 'If in a cross between distinct races or species one sex is abnormal, rare or absent, it will be the heterogametic sex that is so affected.' For it is then ill balanced compared with the other, XX, sex. The latter will have one complete set of genes from each variant, race or species concerned, but in the heterogametic sex is an X chromosome from one parental type but not from the other. This leads on eventually to the situation in which the two distinct chromosome sets are sufficiently dissimilar no longer to act harmoniously in the XX sex also.

Once we find that, in the wild, crosses between two populations give rise to imperfections in the heterogametic sex, the first step to speciation has been taken; it is one that has actually arisen in the laboratory also (Dobzhansky and Pavlovsky, 1966). It is hard indeed to reverse it. Selander and his colleagues used protein variation (p. 63) to examine genetic diversity in two subspecies of mice, *Mus musculus musculus* and *M. m. domesticus* in Jutland. They found that these differed at 32 per cent of the 41 alleles in which by this means they could be compared.

There is a large moth, the Grass Eggar (*Lasiocampa trifolii*) which is common in isolated populations along the south coast of England. Kettlewell crossed the race from Devon, which is dark brown, with one from East Kent where it is putty-coloured. That distinction is unifactorial, the dark shade being dominant to the pale; but he detected something far more interesting than that. The males, XX, were sexually normal while the females, XY, were sexually imperfect: the two races, having evolved 450 km apart, had already taken the first, and decisive, step towards speciation.

We constantly find such disharmony in varying degrees when we cross races and closely related species. It generally leads to complete sterility in the offspring. A scent difference prevents the pairing of the flies *Drosophila pseudoobscura* and *D. persimilis* in natural conditions, but this can be overcome in the laboratory. It is then found that the F1 males (XY) are sterile but the females (XX) are fertile. Though these latter can be back-crossed successfully to males of either parental type, we have here two distinct species.

The evolution of specific differences is generally a gradual process. It can, however, occur at a single step by doubling the chromosomes of a sterile hybrid, so that the dissimilar paternal and maternal chromosome members each gain a partner with which to pair. That, however, is a matter to be described in Chapter 6 in relation to chromosome multiplication. A difference in chromosome number evidently leads to imperfect chromosome pairing, but this can be overcome in a similar way.

What then constitutes a species? That stage has been reached when a form has become sterile, though not necessarily completely so, with any other; also when the hybrids cannot be maintained as a true-breeding race because in at least the heterogametic sex (XY) they are completely or relatively sterile, even when the XX progeny are fertile. Thus, as Mayr has said in 1969, 'Species are groups of interbreeding natural populations that are reproductively isolated from other such groups.' Of course such a definition is very often difficult or impossible to apply in practice: in general, it can only be predicted that such sterility must exist. Evidently the test is sometimes basically inapplicable, as in plants and animals that do not reproduce sexually. A specific status must then be judged on general considerations, as the higher classificatory groups (e.g.

genus, family) have to be. The discovery of full fertility or sterility between groups is, however, a clear indication that they are or are not specifically distinct. It is only fair to add that when such a test has later been applied to populations originally judged as specific or not on mere anatomical grounds, it has usually, though not always, shown that the conclusion so reached has been sound.

Before turning to polymorphism in the next chapter it seems appropriate to think briefly of a highly exceptional, though to us most important, instance of evolution; really unlike any other. It is that of man himself. He is an astonishingly varied species, perhaps to be compared with brambles and wild roses (pp. 127, 137). For he, like they, has developed a large number of 'races' which appear to have remained fully fertile with one another; though how far that is true we have at present no means of knowing (pp. 170–1). But this is a question of the 'recent' adaptations of man as a human being, one to revert to on pp. 174–81. Here we need to think of a stage far more profound. What has enabled the ancestors of man to adopt a wholly unprecedented type of evolution, one based upon increasing intelligence, and on that alone: something indeed which has placed a rather primitive type of mammal, for such in structure man is, at the head of progress? Indeed so great a departure from any other evolutionary course could only be possible in an unspecialized creature not committed to any ordinary types of adaptation.

To follow the line it has taken, mankind has required a slowing down of development, both before and after birth, in order to lengthen babyhood and the period of learning. Indeed the effect is extreme: many mammals are fully active at a period in their lives when man is a heedless infant. In this matter the early, prenatal stages of the process are critical, setting the pace, as well as the direction, as it were.

We can obtain insight on this matter by looking at its converse. I have had the advantage of dissecting several thousand wild mice (voles) over a period of some years, and the work gave me much experience of their embryos throughout gestation. One feature seemed to me particularly striking. In early development there were not only more foetuses than ever were born, but more than could be born, on account of the limited space available.

A little later, one saw that the embryos were growing at different speeds, although they were in so constant an environment as a mammalian uterus, and that the larger began to constrict the smaller and limit their blood supply. Soon the latter were being resorbed, while a few rapid developers were forging ahead. Evidently fast growth was so favoured in these circumstances that the laggards never had a chance to display their qualities, however good in other respects these might be.

This is the direct opposite to the human situation, in which relatively slow growth is essential. How has that been achieved? It results from the fact that man, and the primates most nearly related to him, produce on the whole but one young at a birth, the number of conceptions being under genetic control, so that such embryos are in general free from competition. It is owing to this and because, as Julian Huxley and I showed (1925, 1927), rates of development vary genetically and are therefore open to selection, that the possibilities of human evolution have been realized.

4. Polymorphism and Mimicry

Polymorphism

The types of genetic variation so far discussed are, first, that due to the simple segregation of a pair of allelic genes, and also of two or more pairs, including their independent assortment and the effects of linkage; and, second, the multifactorial type, of which there are examples in Chapter 3. Both are of a kind capable of splitting a population by separating it into groups each with distinct features of its own; occasionally, indeed, up to the evolution of separate races.

We now turn to something fundamentally different in its effect: variation of the reverse kind, holding different forms together. This is known as *genetic polymorphism*. I defined it in 1940 and it will be useful first to state that definition and then to think of its meaning. 'Genetic polymorphism is the occurrence together in the same locality of two or more discontinuous forms of a species in such proportions that even the rarest of them cannot be maintained merely by recurrent mutation.' This definition becomes clear and meaningful when considered piecemeal.

We are here dealing with two or more forms, or 'phases', of one species occurring side by side in the same locality: excluding geographical variation, that is to say. Moreover, these are distinct and contrasted as, for instance, the human blood groups are well known to be (pp. 139–48); thus they are not connected by a series of intermediates, as in human height. In addition, polymorphic variants have no overall disadvantage compared with the normal condition; otherwise they would be eliminated by selection and maintained merely by mutation. Consequently, the human race is not polymorphic for rare diseases due to wholly disadvantageous genes: Huntington's chorea (p. 162), for example. Considering the rarity of mutation, it may indeed be said that any clear-cut form must be polymorphic if it occupies as much as 2 per cent of the population.

It will be noticed that we are making the assumption that polymorphic variation is normally genetic. One of the very few instances in which it is environmental is provided by the subdivision of female bees into queens and workers, brought about by the food supplied to the larvae. A similar situation applies also to the 'caste' differentiation in some other insects. Seasonal variation, though rightly excluded by the definition, might seem to approach polymorphism. An example of it has already been given in the butterfly *Araschnia levana* (p. 33). Such variation is not polymorphic, though it has been claimed as such; for the distinct types do not occur together. Consequently the special adjustments and selective balance for maintaining both at the same time have not had to be evolved.

Evidently to become polymorphic a gene must gain some advantage and begin to spread; but its value must then decline, reach neutrality and pass to a disadvantage if it becomes still commoner. Otherwise it would displace its former normal allele, which would then exist merely as a rare mutant. The polymorphism would then have disappeared.

How can a quality lose an advantage and become harmful as its proportion rises in the population? There are two principal ways, often combined, in which this can happen.

First, two forms can reach balanced frequencies by the very nature of the difference between them. For instance, sex clearly falls within the definition of polymorphism, and is a situation in which neither males nor females must become too common or too rare; indeed they will obviously reach optimum proportions in any species, generally near equality. Similarly, we shall find a balance of advantages between the forms in the Batesian mimicry of butterflies, as we shall see later in this chapter (pp. 86–8).

Second, the genetic control of variation can itself lead to a balance of forms; for this must arise if the heterozygotes are at an advantage compared with their two homozygotes: a condition that ensures permanent variation, since neither of the controlling alleles can then oust the other. It is a situation that tends to evolve whenever a gene proves successful; consequently polymorphism must be very frequent.

Such 'heterozygous advantage' can arise in either of two ways, and one of them is so easy to follow that it can appropriately be mentioned here. A gene has multiple effects and arises by mutation. This is generally harmful since unrelated to the needs of the organism. Thus if one effect be advantageous, the others are very unlikely to be so too. Advantageous qualities tend to become dominant because the individuals in which they have the greatest effect are favoured in the heterozygotes. For a similar reason, disadvantageous qualities tend to become recessive.* By definition (p. 21), the (advantageous) dominants act in the heterozygotes, while the (disadvantageous) recessives do not. Therefore the homozygotes have an advantage

* Dominance and recessiveness are discussed on pp. 99–101.

and some disadvantages, while the heterozygotes have advantages only.

It is now possible to examine instances of polymorphism and to see where they lead us.

One of the commonest British and western European snails, *Cepaea nemoralis,** is highly variable in many parts of its range, though by no means in all (pp. 81–2). The shell-colour may be yellow or dark; pink or brown. These shades are discontinuous; that is to say, one finds few or no intermediates between them. Moreover, the specimens may bear up to five dark bands, or they may be bandless.

Some of these types match the background better than others. To human eyes, the yellow, greenish when the animal is within, are less conspicuous than the pink or brown on short grass, while the reverse is true on dark leaf litter in a wood. Also the banded specimens are the less easy to see on a diversified background, such as a mixed hedgerow, while the unbanded are the better concealed on a uniform one.

This snail is preyed upon by thrushes. The birds swallow the small specimens whole, but they carry the larger ones to favourite stones where they break them open. There the shells accumulate, so that by examining the remains it is possible to decide if the snails are taken at random. This they are not, for the most conspicuous specimens are chiefly destroyed, and those that are the best concealed to human eyes are evidently so to those of birds.

By marking the snails with a dot of cellulose paint†, placed underneath so as to be invisible to the thrushes, it has been shown that the difference in frequency between the predated individuals and those living in the immediate neighbourhood is a real one. That is to say, it is not due to the birds changing their feeding ground and bringing the snails from a distance to be broken on their favourite stones.

In woodland habitats, the comparison between the living specimens and those destroyed by the birds needs to be made with caution. For as the ground cover grows, changing the carpet from predominantly brown in early April to green in mid-May, the proportion of brown shells destroyed increases and that of the yellow declines, in accord with the type that is best concealed as the season advances.

* It is sometimes referred to as the Grove snail or Glade snail, mere translations of the Latin which do not seem very suitable names in English.

† To do so, it is necessary to remove the shiny outer coating of the shell (the periostracum) over a small area, with sandpaper or a file; otherwise the paint will not adhere.

Now do the birds influence the local evolution of this snail? If its colour pattern were environmentally controlled, they would not. True, even then they would tend to adapt a population to its background by removing the more conspicuous specimens; and these are not the same in one habitat as in another. But such adaptation in one generation would not be cumulative by influencing the next; each would have to start with a similar disadvantage. In fact we know that both shell colour and banding are indeed genetic, so that selection acting upon them brings about evolution.

The colouring is controlled by single genes that form a multiple allelic series, the dominance order being brown, pink, yellow; so that the latter is recessive to both the others. Also the unbanded type is dominant to banding of any kind, the various forms of which are due to 'modifying' genes, having no effect on the bandless type. Two or more bands may be fused, and this also is a genetic feature.

It is clear then that in some localities at least, the colour-patterns of *Cepaea nemoralis* prove to be adjusted to suit its habitat. From many places where Cain and Sheppard (1950) studied this snail, they chose five with the most green at ground level and five with the most brown. They found that:

The *lowest* percentage of yellow shells on a *green* background was 41.

The *highest* percentage of yellow shells on a *brown* background was 17.

Taking also the five most uniform habitats and the five most varied, they found that:

The *lowest* percentage of unbanded shells on a *uniform* background was 59.

The *highest* percentage of unbanded shells on a *varied* background was 22.

It is especially to be noticed that though the inappropriately coloured and patterned snails are constantly being eliminated in each habitat, the populations there do not become uniform; a clear indication that the genes for colour and pattern are responsible also for physiological adjustments. These therefore *maintain* the diversity, though the proportions of the forms in each locality are *adjusted* by bird predation.

From the way in which heterozygous advantage must tend necessarily to arise in polymorphism, that condition must be one of its characteristic attributes. Thus the genes affecting the snail's colour-pattern must influence also its physiology. Indeed it has been shown that the different colour and banding types do not withstand unfavourable conditions equally well.

It must not be supposed that visual selection by birds can be observed throughout the range of this snail. There are extensive regions in Britain where its appearance, matching its background, is unimportant compared with the, still polymorphic, physiological effects of the genes for colour or banding, maintained by hetero-zygous advantage. Such 'area effects' as they are called, are espe-cially to be found in dry environments; those for instance on chalk downs and sand dunes. In such places, the situation in Britain somewhat resembles one that is widespread in France, and there the colour-pattern is often unrelated to background. It is a telling fact that archaeological excavations on the Marlborough Downs show that some of the area effects of *C. nemoralis* found there today were already established in neolithic times (say, 2,000–800 B.C. in that region). Here we see how permanent genetic adjustments may be when necessary, though the species can adapt quickly to a change in its habitats.

Polymorphism often leads also to the formation of a 'supergene', and this is another important aspect of it. The complex adaptations of organisms are likely to require the co-operation of many genes. When all of them are acting, they can be scattered among the chromosomes; those for the proper development of the mammalian heart, for example. Not so in polymorphism, in which two or more alternative forms are adjusted to different sets of conditions at the same time. If genetically controlled by genes assorting inde-pendently, incorrect combinations suited to no single environment, would constantly arise. This is largely prevented by bringing the genes in question on to the same chromosome and ensuring that they will seldom be separated by crossing-over. When they thus tend to act as a single unit, they are said to form a *supergene*.

In *Cepaea nemoralis*, the presence or absence of bands often needs to be combined with a particular shell colour. The genetic control of these two features is distinct, for all combinations of them may be found in the countryside. But those desirable in any one habitat are held together because controlled by genes so closely linked that a cross-over seldom occurs between them. So effective is this linkage that for long it seemed absolute. But later two experimental broods were obtained in which crossing-over of 2·25 per cent occurred between the two loci; for cross-over frequencies can vary genetically, and are therefore open to adjustment.

How does a supergene arise? There are several ways in which some linked genes can be held together more or less permanently; but here we need only consider *inversion*: that is to say, when a section of a chromosome has been reversed. Now chromosome pairing, essential for meiosis (p. 26), is due to an attraction between allelic genes. It can take place unimpeded anywhere along a pair of chromosomes that contain no inversions and along a pair con-taining homozygous inversions; for both are situations in which corresponding genes have their loci directly opposite one another. The two types may be represented as:

ABCDEFGHIJKLM----   ABCDHGFEIJKLM----

ABCDEFGHIJKLM---- and ABCDHGFEIJKLM----

Alternatively, such attraction cannot occur between a non-inverted and an inverted region, for there the members of the alleles are no longer directly opposed:

ABCDEFGHIJKLM----

ABCDHGFEIJKLM----

Consequently a cross-over is not effective in producing recombination within a region subject to inversion, which holds its included units together as a supergene.

But what if, as is so likely, the genes which it is desirable to associate start on different chromosomes? They can then be brought on to the same one by means of *structural interchanges*; that is to say, pieces from non-homologous chromosomes break off, and then become reattached but interchanged. This occurrence is rare, but not unduly so. Jacobs (1971) found four instances of it among the chromosomes of 2,538 men, and that is much above mutation frequency.

We have just encountered evidence of heterozygous advantage and the formation of a supergene in a snail. Yet these features are the more particularly clear in the fly *Drosophila pseudoobscura*, abundant in parts of the U.S.A. Its polymorphism is purely physiological. All the forms look alike superficially. They are to be distinguished only microscopically by an examination of their chromosomes, subject to abnormal pairing where there are inversions; these are polymorphic and hold their genes together.

The numerous inversion types in this species are indicated by pairs of capital letters, not italicized so that they cannot be confused with gene notation. When studying two inversions, AR and CH, at given localities and temperatures in California, the distinguished geneticist Dobzhansky (1950) attributed a survival value of unity to the heterozygotes. He then compared the three classes AR/AR, AR/CH, CH/CH and found that their relative survival values were 0·71 : 1 : 0·43. Indeed the heterozygotes of these and other inversions constantly exceeded the survival of the homozygotes. This, moreover, is not merely some outcome of heterozygosity itself but has *evolved*, depending on the gene combinations associated with it. When, as just explained, the AR/CH heterozygotes were superior in their survival to the two homozygotes, the flies came from the same locality. But when they were the result of a cross between individuals caught 1,000 km apart, the advantage of the heterozygotes disappeared and the homozygotes were actually superior to

them, the values being 1·53 : 1 : 1·16. This is due to the fact that
at each place the adaptations involving heterozygous advantage
had depended upon adjustment to local conditions; consequently,
the effects are largely lost on wide out-crossing. The evolutionary
process is here demonstrated by reversing it: a technique that shows
the heterozygotes to have no advantage *as such*.

Quite similar 'cryptic' polymorphism, with heterozygous advantage, is constantly being encountered because it can now be detected
chemically (p. 63). Several such polymorphisms have already
been discovered in the Meadow Brown butterfly, and reference is
made to two of them at the end of the previous chapter. These
show marked instability across the 'boundary area' (pp. 71–2)
extending from mid-Dorset to east Cornwall, in which they confirm
on very different grounds the unusual response of that insect to this
particular tract of country, elicited by a study of its spot numbers.

Indeed polymorphism of this kind is widespread, and there is
much evidence to show that it normally involves heterozygous
advantage. It occurs, of course, in plants as well as animals. For
example, two enzyme systems are polymorphic, at different frequencies, in two species of oats.

It is obvious enough that a feature harmful in one region and
maintained there merely as a rare mutant may be polymorphic in
another because advantageous in the environment there, whether
external or genetic. We shall find telling instances of this among the
human blood groups in Chapter 7. Indeed the special local conditions enabling a normally unsatisfactory condition to spread are
sometimes apparent. For instance 'frizzle' in domestic poultry is an
incomplete dominant. The feathers are imperfectly formed and
turn up at the ends; therefore they fail to provide effective heat
insulation, and the birds survive only if protected in cold weather.
This is the heterozygous state. The homozygotes are far more
severely affected. They seem of poor stamina in any circumstances
and must be maintained in an incubator, for the feathers are sparse
and grossly abnormal. Yet I found that perhaps a quarter of the
poultry round the native villages in Fiji were of the 'frizzle' type,
capacity for heat loss being a positive advantage in that climate, to
be set against the fact that the homozygotes were still unable to
survive even there; consequently a polymorphism had arisen for this
quality.

Biological Mimicry

Polymorphism will be referred to many times in the succeeding chapters, as will additional aspects of its evolution, especially in regard to industrial conditions. However, some of its chief features may appropriately be further illustrated now by means of biological mimicry. This is ideally of two kinds, Mullerian and Batesian;* though, as we shall see, they merge into one another and can be combined in different degrees. Though found in other groups of animals, and indeed in plants, they are particularly well shown in butterflies and day-flying moths, since the colours and patterns of these insects give so much opportunity for recognizing their variation, and because of the bird predation to which they are subject.

Certain of the species are extremely distasteful, while others are poisonous; even a droplet of the body fluid producing nausea and vomiting in birds, owing frequently to the heart poisons† that it may contain; effects that even traces of these substances produce. The Hon. Miriam Rothschild and her colleagues (Rothschild, 1972; Rothschild and Reichstein, 1976) have shown that these chemicals are widespread in eleven families of plants, and that they can be stored in the bodies of butterflies and moths whose caterpillars feed upon such species (e.g. the Monarch, *Danaus plexippus*, and the Polka-dot moth, *Syntomeida epilais*).

Though many butterflies and moths derive their repellent qualities from their larval food, others manufacture disagreeable substances and poisons themselves, as do the Burnet moths (these synthesize acetylcholine, histamine and hydrocyanic acid). Such 'protected' species therefore evolve a striking appearance, easily recognized, by means of bright colours and a simple pattern and, often, slow flight. They are also tough and leathery, and thus able to withstand minor injuries from pecking, when they exude a droplet of their unpleasant body fluid. Young birds have to learn by experience which insects to avoid on account of these qualities. The distinctive appearance associated with poisonousness is termed a_____ *ic* (warning).

_____ several such species to resemble one another, so that _____ inedibility serves for all. This is *Mullerian mimicry*, _y is much enhanced by the long memory of birds, _oths. As such adaptations aim at securing uniformity _, we generally look in vain for polymorphism here. _urnet moths, just mentioned, are strikingly aposematic. 5 cm or a little more across the expanded wings. The

_____ which commemorate F. Muller (1831–97) and H. W. Bates
_____ who first recognized them.
† Calotropin and calactin.

species, of which there are numbers, are all metallic green or blue-black, with remarkably similar scarlet markings. They are widespread in the Old World, though absent from America. Only one among them, *Zygaena ephialtes*, is polymorphic. In addition to the characteristic colouring of the group it has a widely different alternative form, black with white spots, a feature explained by its exceptional ecology. Rothschild finds that a mouse is killed in two or three minutes when the body fluid of a single female of the European and southern English Burnet moth, *Zygaena lonicerae*, is injected into it; the female is the more deadly since it contains a higher concentration of the poison, with which to protect its eggs. It seems that this situation is common. Rothschild has shown that the conspicuous caterpillars and the adult males of the Scarlet Tiger moth, *Panaxia dominula*, can be injected into a mouse without ill effect, while a similar extract of the adult female proves fatal in twelve to fourteen hours.

Many butterflies are mimetic in the tropics and subtropics of the Old World and the New. The concentration of that condition there is due to the intense competition of the teeming life in those regions, especially in the forests.

Butterflies, moths and other insects generally attempt to escape their enemies by concealment or by a fast and erratic flight. Yet some gain protection by copying species that are distasteful. These become their 'models', and *Batesian mimicry* consists in the one-sided advantage of resembling them. Evidently such 'mimics' must not become so common compared with their models that a given, and necessarily conspicuous, colour-pattern tends on the whole to indicate something edible rather than the reverse. The result is that, unlike the Mullerian situation, this numerical limitation can be, and often is, mitigated by polymorphism. The mimic then copies as accurately as possible several models, and it must do so in a clear-cut fashion, of course without intermediates. Here then we meet the paradox of extreme *diversity* involving highly distinct forms, each of which must be *constant* in appearance. We should therefore expect, and indeed we find, them to be controlled by single genes or, much more usually, by supergenes.

It will be noted also, as Rothschild (1971) was the first to point out in her important survey of mimicry, that polymorphism may avert the danger of small numbers. For extreme numerical reduction can lead to at least local extinction in a highly unfavourable season.

A few general features of Batesian mimicry can now be mentioned. In the first place, model and mimic need not belong to closely related groups: it is only necessary for them to look alike superficially.

Thus their similar colours may be produced by chemically distinct pigments, while apparently corresponding patterns can occupy different positions on the body or wings.

The reality of mimicry has repeatedly been demonstrated experimentally, and in a number of ways. Birds brought up in captivity may find a mimic highly edible until supplied with its model, which they will not have seen. This they mistake for it and discover it to be disgusting. Subsequently they refuse its mimic, which they had previously accepted. In another experiment, Dr. Jane Brower coloured some mealworms with a green band of paint and offered them to insect-eating birds, starlings, which eat them voraciously. These were then made unpalatable by dipping them into quinine solution. After an initial trial, the birds would avoid the green-banded mealworms whether they had been dipped into quinine, so becoming 'models', or not; the latter being 'mimics'. She found that most (80 per cent) of the mealworms marked with green were protected from predation if only 30 per cent of them or even less had been made distasteful. It is, therefore, by no means necessary for the models, if decidedly unpleasant, to outnumber the mimics.

Batesian but not Mullerian mimicry is often confined to the females, the different polymorphic forms of which though autosomal are then sex-controlled (pp. 49–50). This, as Rothschild has pointed out, enables the mimic to be twice as common as it would otherwise be. The females, moreover, are in the greater need of protection, for they expose themselves when egg-laying; while a relatively large destruction of males can be tolerated, since a single one can generally fertilize several females. Furthermore, a situation which actually favours male uniformity arises from the fact that the stimulus required to induce a female to copulate is often visual, depending on the appearance of her partner. This sometimes extends further than the human eye can see, involving markings evident only in ultra-violet light, visible to some butterflies but invisible to man. We need to know more about this. Are such colours involved in mimicry? Which predators see into the ultra-violet and which do not? There appears to be diversity here. Many birds seem deficient at the blue end of the spectrum, but some (e.g. the bower bird, *Ptilonorhynchus violaceus*) shows a preference for it. Is there then, in some instances, a correspondence between an ultra-violet pattern in model and mimic; and what about its variability?

Polymorphic mimicry frequently includes non-mimetic forms in addition. The latter are maintained by heterozygous advantage, as in polymorphic species that are not mimetic at all: the brown (recessive) and the blackish-green (dominant) females of the Silver-washed Fritillary butterfly, *Argynnis paphia*, for instance. The frequencies of the phases, mimetic and non-mimetic, are then adjusted until all obtain equal advantages.

Three fundamental features of polymorphism can now be illustrated from Batesian mimicry: gradual evolution within the control of a single gene or supergene, switching on alternative forms; the fact that heterozygous advantage evolves under the cloak of mimicry, just as it evolves to produce non-mimetic polymorphism; and the control of polymorphism by means of a supergene.

The large African Swallow-tail butterfly *Papilio dardanus* is widespread south of the Sahara. The males are always tailed, like the species related to them. They are yellow with black markings above, and are non-mimetic. The females are, in general, utterly different. Normally they are tailless, owing to the action of a single autosomal gene when homozygous, affecting the females only. Tails are present but short in the heterozygotes, while they are of full length in the other homozygote.

The females exist in a number of polymorphic forms. About five of these copy distinct species known to be poisonous or distasteful, though sometimes in part of their range only, and in this they are so successful that the males are frequently deceived until they make a close approach. These mimics, so distinct from one another, are yet very constant in appearance, while the non-mimetic forms, of which there are several, are variable.

Clarke and Sheppard (1960) have studied the genetics of this situation with great success. They find that the polymorphic females are due apparently to a series of multiple alleles acting in that sex only and situated at a locus distinct from the one controlling the presence or absence of tails.

An obvious difficulty has to be faced at this point. The resemblance of mimic to model involves colour, pattern and habits (whether tending to fly in the sun or the shade, and with a gliding or a flapping flight and, probably in some instances, scent). Yet the gene or genes concerned must have arisen suddenly by mutation. But it is incredible that the species has had to wait for the chance occurrence producing the correct effect in its diverse manifestations and perfection. In fact the mimetic features have been evolved gradually, from a slight accidental resemblance, by selection acting upon genetic variability within the effect of the controlling gene. That view has now been fully confirmed.

For instance, *Papilio dardanus* occurs also in Madagascar where, however, the females are tailed and not only non-mimetic but male-like. When a mimetic female from South Africa is crossed with a Madagascan male, the effect of its gene for mimicry can be studied in a genetic background to which it is not adjusted. It is then found that the accurate mimicry breaks down. Yet it can gradually be restored by successive back-crosses to the mainland race. This therefore must carry genes for perfecting the mimicry within the working of the gene or supergene that controls it. Thus Clarke and Sheppard were able to unravel these polymorphic adaptations by, as it were, taking them to pieces and putting them together again.

A further aspect of polymorphism must be noticed here. Though *Papilio dardanus* occurs quite commonly in the mountains east of Lake Victoria, being polymorphic in the females, yet its models are nearly absent there. Thus mimicry is not necessary for the survival of the species. On the other hand, there is normally selection for it, since in this exceptional region the resemblance is no longer accurate and the mimetic forms are variable. That is to say, in this habitat the polymorphism must be preserved by its heterozygous advantage alone, as it is elsewhere when non-mimetic; the situation has indeed much in common with that giving rise to 'area effects' in the snail *Cepaea nemoralis*.

Though the genes controlling the polymorphism of *P. dardanus* behave as if they were multiple alleles, occasional cross-overs between them have been detected, and indeed provide the genesis of several of its rare non-mimetic phases. Evidently, therefore, these apparent multiple alleles are in fact supergenic. The original controlling gene may not have been that affecting colour or pattern.

It has been mentioned that Batesian and Mullerian mimicry, which originally appeared so distinct, can to some extent merge into one another. That consideration can be illustrated by a remarkable discovery, again due to The Hon. Miriam Rothschild and her colleagues. She finds that the similarity in appearance between the ordinary White butterflies (Pieridae), the Cabbage Whites and their allies, is due to mimicry in which Mullerian and Batesian features are to some extent combined.

The Large White butterfly, *Pieris brassicae*, a pest of cabbages in Europe including Britain, is throughout its whole life history protected by an unpleasant taste and poisonous qualities to an extent far exceeding the species related to it. Marsh and Rothschild (1974) find that a single female, ground up in saline, will kill a mouse in 30 hours when injected into its body cavity. The male is less lethal, causing death in four days; while the chrysalis is more so, for the death of a mouse follows such an injection in 10 to 15 hours.

Powerful protection of this kind needs to be made apparent so that it can be recognized by potential enemies. The eggs are yellow in spite of their green background, and laid in clusters on the cabbage leaves, without concealment. The caterpillars are gregarious and do not at all match their food plant in colouring, while the chrysalis is most unusual. In a number of the White butterflies its colour is adjusted to resemble its background (a situation which I have illustrated (1971, Plate 2). That of the Large White is also adjusted in regard to its background but in the opposite way: to produce a contrast, so as to make it easily visible in all circumstances. The Hon. Miriam Rothschild showed that this remarkable adaptation proves to be possible only in the presence of at least a small quantity of a carotenoid pigment that is derived from the larval food and chemically necessary for this particular reaction. The adult butterfly itself is highly conspicuous, while Rothschild has shown that birds remember for months the unpleasant experience of trying to eat it.

The Small White, *Pieris rapae*, being often mistaken on the wing for the Large White, gains protection from that resemblance. Yet it, too, is unpalatable, though not to such an extreme degree: about half. This serves the active butterfly in good stead, considering that the Large White is much the less abundant, and makes it also a species worth copying. Thus the Small White acts on the one hand as a (Batesian) mimic and on the other as a model. Its early stages, however, gain such security as they possess by concealment. The eggs are green and solitary; so are the caterpillars which, in addition, hide by burrowing into the cabbage plants; while the chrysalis adjusts its colour to *match* its background.

Why do not the early stages rely upon advertising the unpleasant qualities which the adults certainly possess? As Rothschild has pointed out, the egg, caterpillar and chrysalis must all be exceptionally distasteful, as are those of the Large White, if they are to depend upon this alone for protection, since they cannot escape by flight. It is not surprising, therefore, to find instances in which caterpillars may acquire a violently unpleasant taste, as those of the Lackey Moth, *Malacosoma neustria*, when feeding on laurel, though the resulting adults are regarded by birds as a choice morsel.

We find also Batesian mimicry, with the Large and Small Whites as models, in a number of related but palatable species, for instance the Green-veined White, *Pieris napi*; and the female Orange-tip, *Anthocharis cardamines*, but not its orange-marked male. And it will be remembered that mimics are often confined to the female sex.

A few other generalizations can now be made in regard to mimicry. They are derived largely from the penetrating studies of The Hon. Miriam Rothschild, which have contributed so greatly to our knowledge of this subject.

Birds tend to pick upon the largest prey available. Consequently mimics are generally smaller than their models; the Large White

is the largest, as well as the best protected, of the species just mentioned. That situation may, however, be modified in polymorphic
forms, the numbers of which tend to be so adjusted that all enjoy
equal advantages: thus the polymorphic females of *Papilio dardanus*
are larger than their models. As a result, primarily of Rothschild's
work, we now know that by no means the whole population of
certain models is protected by poisonous or distasteful qualities; in
some areas she has found diversity in this respect. Moreover, she
has shown that it may be possible for a Batesian mimic to acquire
repellent attributes by feeding upon unusual plants.

Thus Mullerian and Batesian mimicry can merge into one another.
Yet, on the average, the distinction between them has reality; as is
shown by polymorphism, for that is almost entirely restricted to the
Batesian situation. It is hardly known in the Mullerian one, which
exists to promote uniformity.

Rothschild has suggested that aposematic species (p. 85) may
have an 'inverted' advantage: being, from one point of view, useful
to their food plants. For upon these they tend to sit conspicuously
for long periods, attracting to the flowers additional insects which
can also act as pollinators. In Chapter 6 we shall consider a balanced
interrelation between plant varieties manufacturing, and destitute
of, poisons protecting them from herbivorous vertebrates.

The exceptional features of mimicry, particularly that of butterflies and day-flying moths, for long made its evolution obscure. In
recent years, work on ecological genetics and chemistry has provided a decisive explanation of it in terms of selection: an advance
which will always be associated with the names of Sir Cyril Clarke,
The Hon. Miriam Rothschild and the late Professor P. M. Sheppard.

Finally, it may be said that an attempt has been made by Cavalli-
Sforza and Bodmer (1971) to replace the definition of polymorphism
given on p. 78 with one depending on 'the occurrence of two or
more alleles at one locus, each with appreciable frequency'; this is to
avoid an inherent assumption of selection. How such multiple
alleles have spread until each reaches 'an appreciable frequency' is
not apparent (p. 56). Moreover, such a definition is contrary to
what is in general an ascertained fact: that what appear to be
two or more alleles at one locus are in general not so, but are
closely linked genes. As a study of this chapter will show, the
definition of Cavalli-Sforza and Bodmer is a failure precisely
because it aims at excluding selection rather than providing a
concrete epitome of polymorphism.

5. *Industrial Pollution*

One serious effect of industry in any part of the world is the pollution of the countryside to which it so often leads. This may be airborne or waterborne, or it may take the form of masses of unusable, sometimes poisonous, waste material. We have here a problem that, in its various aspects, has a two-way impact on genetics.

On the one hand, genetic studies may help to combat this nuisance and potential danger. For they may provide a sensitive test of such contamination, so that it can be recognized and checked at an early stage, and may thus to some extent overcome the disastrous consequences involved. From another point of view, the theoretical, the effects of industry can so alter the environment as to produce a greater change of habitat than that encountered when a species is transported to a distant land. They may therefore lead to such rapid adaptation that the progress of important evolutionary adjustments can be followed not only in the laboratory but in the countryside. Considered in that light, this chapter provides a valuable extension of the two previous ones on evolution.

Pollution by Smoke

Since the middle of the last century, more than a hundred species of moths have turned black in certain areas of Britain. That is to say, a considerable proportion of the population in those places has done so; and this has occurred in a few insects belonging to other groups also. The process has generally originated in manufacturing districts, as the effect of air pollution; and the black specimens have sometimes spread thence in decreasing frequencies far into apparently uncontaminated country. Though recognized first in England, a similar transformation has taken place in Germany, Czechoslovakia, the U.S.A. and other regions.

It will be helpful first to discuss this event in a single species in which it has been studied extensively; and then to compare it with other instances, also well documented, which add to our knowledge of the subject in various ways.

The Peppered moth, *Biston betularia*, is about 5 cm across the expanded wings. It is common throughout the mainland of Britain, the caterpillars feeding on a wide variety of trees and shrubs. The wings are whitish, marked with a scatter of minute black speckles and thin lines. It has one generation in the year, flying from mid-May to early July. Its habit is to rest upon tree trunks and posts, wings outspread, when it bears a remarkable resemblance to a patch of lichen.

On the other hand, the form *carbonaria* is wholly black, looking as if dipped in ink except for a white pinpoint at the base of each fore wing. First observed in Manchester in 1848, it rapidly increased in frequency until in that area 98 per cent had become black by 1895. A similar occurrence began a little later in other industrial districts also. In such places the tree trunks are blackened by soot; and lichens, which are highly susceptible to contamination, have disappeared. In such conditions, the normal pale specimens, ordinarily so well concealed, are extremely obvious while the black, though very evident on pale bark, are inconspicuous.

This is of much importance to the moth, since it is preyed upon by insect-eating birds. The cryptic advantages of the two types on their respective backgrounds have been studied in detail; originally by Kettlewell (1973), who had bred a stock of several thousand pupae for the purpose. He carried out his work in two ways: first, by direct observation in unpolluted woods in Dorset. There he released equal numbers of pale and black specimens on to trees, about fifty a day, replacing them when all the specimens of one colour had disappeared, and kept watch on them with binoculars from a hide. In all, 190 specimens were taken by birds and, though the two forms were in equality, 164 of these were black and twenty-six were typicals.

In the same unpolluted region, he also liberated 800 of the moths, distinguishing them from the wild specimens occurring locally by a dot of waterproof paint placed underneath a fore wing and so invisible to predators. He recaptured them subsequently in traps of two kinds which proved equally effective: one illuminated, for the species is attracted to a light; the other bated with virgin females, which produce a scent by which the males find them. Kettlewell's results showed that about twice as many of the pale as the black specimens survived to be caught.

He then undertook corresponding studies in an area near Birmingham where the trees were blackened by soot and the lichens had been killed. His Dorset findings were almost exactly reversed there, for it was the pale typical moths which were predominantly eliminated.

Thus cryptic coloration seems to provide a complete explanation of the spread of the black phase in industrial areas and its absence in genuinely rural ones. But, like so much in science, detailed study showed the situation to be more complex than at first supposed; for in this matter another feature is acting also; that is to say, the three genotypes are not equally hardy. Black coloration is due to a single gene dominant in effect, yet the (black) heterozygotes survive better than their homozygotes; better also than the normal recessives. Actually, the fitness of the pale moths is about half, and of the black homozygotes about 92 per cent, that of the heterozygotes: a situation seen clearly in their ability to withstand unfavourable conditions. It will be evident that a polymorphism has developed here and that, as expected, heterozygous advantage has become established in it.

Consequently the *carbonaria* form will never completely replace the pale one in industrial regions, however much protection its appearance may give it. That is why it has not involved more than 98 per cent of the population of Manchester, although it had reached that level in fifty years. At that frequency 24 per cent of the individuals would still be heterozygotes even if the three genotypes were equally viable. Owing to their physiological advantage, the proportion of heterozygotes must in fact be higher than this.

It is evident that the population of the Peppered moth has *evolved* in industrial areas where it is well established, since it has become predominantly black there. In addition, the high frequency it has reached in these places has made it possible for the insect itself to become adjusted, and in two ways. The early specimens of *carbonaria*, a hundred years or so old, preserved in collections, differ from those found today, for they are not completely black: they bear a slight scatter of light markings. It seems that selection has acted on the gene-complex to favour a genetic environment that intensifies the effect of the controlling gene. This possibility has been examined experimentally by Kettlewell, who has for a number of generations crossed the black specimens found today with pale typical recessives from Cornwall, where *carbonaria* has never been seen. He thus brought its gene back into an unadjusted gene-complex, and did indeed produce black specimens with a slight powdering of white scales on them.

There is also evidence that, as one would expect, the heterozygous advantage of *carbonaria* has developed gradually. This must be true generally, not only in forms adjusted to pollution; though experiments demonstrating the fact have seldom been carried out so far. However, an excellent example of them is provided by the work of Dobzhansky on the fly *Drosophila pseudoobscura*, mentioned on p. 83.

It is interesting to notice that the black form of the Peppered moth has in some places spread far into the unpolluted countryside. This, no doubt, is partly due to the effect of drifting smoke, and of that we have clear evidence; for the increasing frequency of the black form has been reversed in areas where a 'smokeless zone' has been introduced. Thus at Caldy, Cheshire, the proportion of *carbonaria* has dropped from 93·3 per cent in 1959 to 90·2 in 1965. Owing to the large numbers on which the observations were made, the change is significant (it could be obtained by chance less than 1 in 100 times). A similar change in the Manchester area is still more so, with a chance of error amounting only to 1 in 10,000.

At this point, we again meet an unexpected complication. D. R. Lees and others undertook a mathematical analysis of the features involved in the spread of black Peppered moths and found that other agencies are responsible for this in addition to soot. Two of these, of special relevance, may be mentioned here: first, the presence of sulphur dioxide, which is a widespread component of air pollution; and second, mid-January temperature, for the larvae heterozygous for the *carbonaria* gene seem to withstand cold better than those of the other two genotypes.

The importance of smoke in favouring this form of the moth is nevertheless attested by the fact that the amount of sulphur dioxide in the air is but little affected in smokeless zones; yet, as already indicated, the frequency of *carbonaria* is reduced in them. On the other hand, it has reached 70 per cent in East Anglia. One fact suggests that drifting smoke is not responsible for its high frequency there. That is to say, lichens are common in that district and, though susceptible to all forms of pollution, they can withstand a trace of sulphur dioxide in the atmosphere far better than the deposition of soot upon them. It seems, then, that the black form of the Peppered moth can overcome both these deleterious conditions.

It is noteworthy that other black forms of *Biston betularia* are known in addition to *carbonaria* which, though the most extreme of all can occur in much higher proportions than they. These less intensely darkened individuals are collectively named *insularia*. They look as if they were heterozygotes of *carbonaria*, which they are not, being in fact controlled as multiple alleles of it. At least three of them have been identified. One is only slightly darkened and can be difficult to distinguish from the typical insect; one is clearly intermediate in colouring; and one may closely approach *carbonaria* to which all three are recessive, though they are dominant to the normal pale form.

The *insularia* forms play the role of John the Baptist to herald the

D

spread of *carbonaria*, in the presence of which they cannot be detected. Thus it is not possible to estimate their ultimate frequencies, but they are not known to exceed 78 per cent of the population (and this in Denmark). They gain the additional asset of heterozygous advantage, though opposed by some mortality of their homozygotes.

At a moderate level of pollution, *Pleurococcus** and crustose but not branched lichens can survive, and in such circumstances *insularia* is well concealed. Not so when these have disappeared, for it is then that *carbonaria* is so strongly favoured on the bare and smoke-begrimed tree trunks.

We can here usefully consider the occurrence of blackening in a related species. This is the Pale Brindled Beauty moth, *Phigalia pedaria*. It belongs to the same family, the Selidosemidae, as the Peppered moth and, like it, is common throughout Britain; but it appears earlier, having one generation in the year, from January to March. Only the males can fly; the wings of the females are mere vestiges 2 to 4 mm long. Flight then distributes the genes of this insect but cannot extend its range.

When resting on tree trunks or fences, the males are well concealed by their colour pattern, though it is darker than the Peppered moth, from which it differs also in being polymorphic for black forms in the normal countryside. There are two of these: the intensely black *monacharia* and the so-called 'intermediates', in which the blackening is less extreme so that they retain a trace of the ordinary pattern. They are controlled genetically as multiple alleles. The normal (paler) form is recessive to both the black ones, while *monacharia* is dominant also to the 'intermediates'. In these latter, the homozygotes are nearly lethal, for in crosses with the normal form a 2 : 1 ratio is obtained when a 3 : 1 ratio is expected. *Monacharia* shows no such homozygous lethality, though it has developed heterozygous advantage. On the whole, the 'intermediates' of *P. pedaria* play a somewhat similar part to that taken by *insularia* in *B. betularia*.

The two melanic (black) forms of the Pale Brindled beauty taken together amount to 25 per cent or so of the population throughout rural Britain; and they provide a basis for adaptation to pollution, since they become much commoner, up to 75 per cent, in manufacturing districts. It seems that the intermediates never rise above 15 per cent of the total, even in the most industrialized areas.

Lees (1974) has studied the distribution of the three forms of *P. pedaria* in eighty-seven localities in Britain. As with *B. betularia*,

---

* A unicellular plant responsible for the green powdering that often covers bark and palings.

he has analysed it mathematically also: in relation to fourteen environmental features, by means of multiple regression. The most important among them proved to be bark reflectance, which depends on the extent to which the bark is broken up by cracks and upon the presence of lichens, and is therefore reduced when these are killed by smoke; which also darkens tree trunks and other surfaces. Sulphur dioxide, which is so important a component of atmospheric pollution in manufacturing regions, does not promote the occurrence of black forms of *P. pedaria*. In that respect, therefore, the species differs from *B. betularia*.

It seems that the rural melanism of *P. pedaria* is due to heterozygous advantage. A balance is here struck with its loss of cryptic colouring, which must indeed be far less in this species, dark as it is even in its palest form compared with the whitish *B. betularia*. The black forms of *P. pedaria*, of course, actually become an asset in industrial districts.

The effects of air pollution can be analysed further with respect to a beetle, the Two-spot ladybird, *Adalia bipunctata*. Here we encounter a situation basically different from that in the two species so far mentioned, or in any of the large number of moths which have become black in response to air pollution in Britain or elsewhere. For in *A. bipunctata* alone the adjustment has occurred in a powerfully 'protected' insect: that is to say, one completely exempt from predation because of a disgusting taste and smell (p. 85). To exploit this, it makes itself conspicuous and easily recognized by means of 'warning coloration'.

Normally this beetle is scarlet with one black spot on each elytron.* Several polymorphic forms exist, but only two need be mentioned here: *sexpustulata*, black with three red spots on each elytron; and *quadrimaculata*, black with two red spots on each elytron. The latter is dominant to the others, while the normal scarlet is recessive to both the black forms. The three are controlled as multiple alleles.

The adaptations of this beetle to air pollution have been studied by Creed in 140 localities in England and Wales. He has shown that in general the black specimens are quite rare, 2 per cent or less of the population. However, they are common in certain industrial regions, exceeding 90 per cent round Manchester.

Creed finds that the black forms become frequent only when smoke pollution is intense and the coal is of high volatility. The chief source of low-volatility coal in Britain is South Wales; and there,

---

* The elytra of a beetle are the horny wing covers. These are the modified fore wings. The hind pair, concealed and protected by them, are alone used for the motive power in flight.

even in conditions of severe pollution, black specimens are almost absent. He has obtained evidence also that their two distinct forms are differently adapted. *Sexpustulata* is about twice as common as *quadrimaculata* where the local coal is highly volatile and weakly caking, as round Birmingham and in Scotland; while the frequencies of the two are reversed where the coal is also highly volatile but strongly caking. It is so in the Manchester district and, to a somewhat less degree, in Durham.

It has been mentioned that black *Adalia bipunctata* are practically absent from South Wales. To this Creed (1974) finds one exception. They are numerous in the immediate neighbourhood of the National Smokeless Fuels' Phurnacite plant at Abercwmboi. They reached 53·7 per cent of a sample caught 300 m eastwards of it and 41·5 per cent at approximately the same distance to the west. They become rare with increasing distance from the ovens, and are practically non-existent 10 km away from them.* Creed produces evidence to show that the high frequency of the black forms of this beetle near the Phurnacite works is not the result of smoke, and is due to gases other than sulphur dioxide. The subject requires, and deserves, further study.

It has been shown that in the neighbourhood of Berlin, black *A. bipunctata* survive the winter relatively badly and the summer relatively well (their frequencies were 37·4 per cent in April and 58·7 per cent in October; based on several thousand each, the results being consistent over a number of years). There is little evidence of any such effect in Britain. However, it has been shown that the black can survive well at a temperature high enough to kill the red; and Creed suggests that heat adaptation may be more important in the considerable temperature range of a Continental climate than in the more equable Atlantic conditions of the British Isles.

Even the few insects so far discussed in this section are affected differently by distinct components of aerial pollution. We may hope therefore that a wide study among the large number that display polymorphic blackening (over a hundred in Britain alone) may produce a repertoire of species capable of detecting at an early stage different types of such contamination. This may be of considerable medical significance in view of the bronchitis and lung cancer with which such conditions are associated; their importance in that respect has been stressed in a leading article in the *British Medical Journal* (1970, *4*, 256).

---

* There is, in fact, a small unexpected increase in the frequency of the black forms at Ty'n-y-nant, to 10·7 per cent, 15·1 km away from the Phurnacite plant.

It is interesting to consider a little further the genetic and evolutionary significance of such 'industrial melanism', as it is called. Butterflies are not affected by it, for they do not derive protection from sitting exposed on a background which they match in colour and pattern. It is indeed only those species which do so, perhaps to some extent hidden in crevices, that display polymorphic blackening. Such insects as conceal themselves out of sight among vegetation or dead leaves have never been thus affected. It may be noticed that some moths have been able to return as black forms, evolved elsewhere, to industrial districts where they had long ago become extinct: of this, the Rosy Minor, *Procus literosa*, around Sheffield is an example.

One further point must be mentioned here in regard to the darkening of moths in manufacturing districts: that is to say, the effect is not always polymorphic but may be multifactorial. In such circumstances, the whole population of a species in a polluted area gradually darkens. Presumably an appropriate gene capable, furthermore, of conferring heterozygous advantage, has not arisen by mutation in such places. For instance, an intensely black polymorphic form of the Scalloped Hazel moth, *Gonodontis bidentata*, is well known at a high frequency in the Liverpool and Manchester area. This insect has extremely poor powers of dispersal: the males fly little and for short distances only; the females not at all, though they possess wings. Consequently, though the black phase is well established only 30 km or so north of Birmingham, it seems never to have occurred in that great industrial region where, however, the species is common and the whole population is considerably darker than usual owing to selective adjustment acting on multifactorial variation. This same effect is often to be observed also in the normal form of other species that are polymorphic for extreme blackening.

It will at this point be helpful to use the adjustments of organisms to smoke pollution in such a way as to explain and illustrate certain features of general genetics. First we may consider the phenomenon of dominance and recessiveness. These unexpected qualities have constantly been encountered from the earliest days of genetics: the very names were coined by Mendel himself. In examining the dark varieties of the Common Buff Ermine moth, *Spilosoma lutea* (pp. 18–19), we found that the presence of two genes for darkening (in the homozygote) has approximately twice the effect on one (in the heterozygotes). That is what one might expect; but it is what one rarely sees.

At first various explanations of dominance and recessiveness were proposed, mostly, as soon appeared, contrary to the plain facts.

Indeed it was suggested, and repeated almost to the present day, that 'dominant genes' and 'recessive genes' differ in some structural way from one another. Quite the contrary: for the first thing to make clear is that there are no such things. It is not the genes, but the qualities they control, which are dominant or recessive. Thus a single gene can have several effects, some dominant and some recessive and some having heterozygotes intermediate between the two homozygotes. In addition, it is possible to convert a dominant quality into a recessive one, or the reverse, by selection acting on the gene-complex; and to show that the gene itself has not changed: it is the response of the organism to it that has done so.

It was Sir Ronald Fisher who resolved this situation. He pointed out that selection will favour the advantageous effects of a gene so as to give the heterozygotes the benefit of them, and to ensure that those in which they have some effect will contribute more to posterity than the organisms in which their influence is smaller or non-existent. The reverse process will ensure that disadvantageous effects are reduced and obliterated in the heterozygotes, becoming recessive.

It is indeed probable that when a mutant arises for the very first time, two of the genes so produced will generally have twice the effect of one. But since mutation is recurrent and extremely rare at any one locus, such a situation may never have been encountered. It seems that the adjustment which follows operates on the gene-complex to *evolve* the dominance of advantageous qualities and the recessiveness of harmful ones. If that conclusion be true, full dominance or recessiveness should be lost if we put a gene into a gene-complex which has never had a chance of adjusting to it. The species mentioned in the present section provide instructive instances of this.

As might be expected, almost all successful black polymorphic forms are dominants. Recessive black varieties are well known as great rarities in many species. In these they are strongly disadvantageous, and held down to the mutation rate of the gene responsible for them. But it would be strange if just occasionally they did not prove of use in the wholly new type of environment generated by a polluted atmosphere. Three instances of the kind are known.

One is found in the Brindled Beauty moth, *Lycia hirtaria*, a coal-black form of which has reached 83 per cent of the population in the squares and tree-lined streets of London, but not elsewhere. It has been crossed with the Belted Beauty moth, *Nyssia zonaria*, which occurs abundantly but very locally, and only on the coast of Wales, the Scottish islands and the west coast of Ireland; in it black forms are unknown. It never normally crosses with *L. hirtaria*; but it can be made to do so in captivity, though with difficulty. Using first normal specimens of the latter, the hybrids were intermediate in colour pattern between the two and showed no trace of darkening. When, however, the *L. hirtaria* parent was black, the offspring of the species-cross were dark, though they carried but one gene for that shade; yet a pair of genes is normally required to evoke it, for the melanism is fully recessive: a clear example of the fact that the recessive colouring is not due to 'genes for recessive melanism' but to a gene-complex so adjusted as to annul the effect of the controlling gene except in double dose as shown by C. J. Cadbury.

Dr. H. B. D. Kettlewell provided a perfect demonstration of the way in which, on the other hand, dominance can break down in an unaccustomed genetic background. In the *carbonaria* form of *Biston betularia*, the heterozygotes and homozygotes are both black, and they are indistinguishable. He crossed this species with a closely related north American one, *B. cognataria*, which indeed has itself a dominant black polymorphic phase. However, the dominance of the two had been evolved independently in the Old World and the New. In the hybrid offspring, the dominance of *carbonaria* completely broke down and gave rise to a continuous range from pale coloration to black. Clearly then the melanic effect had *evolved*, through selection acting upon the gene-complex.

The fact that dominance and recessiveness are the result of selection leads to a number of important consequences, two of which can be mentioned at this point. Members of a multiple allelic series having a recessive effect will be disadvantageous (that is why they are recessive) and so ultimately maintained merely by mutation. Each therefore will be a rarity, slowly eliminated. It follows that the chance of combining two of them together in natural conditions must be remote. Thus there is virtually no opportunity for selection to act upon such a combination, so that its effect will remain intermediate between that of each allele concerned; yet these will probably have achieved recessiveness with the normal form.

When a gene is very harmful indeed, the individual carrying it is generally eliminated without leaving descendants. There is then no opportunity for selective adjustment, so that the condition remains intermediate. This is the situation found in many of the more dangerous human diseases when controlled by single genes.

We can now turn to another general property of genetics to be illustrated from the effects of air pollution. It has been said that in the hundred and more species of moths that have developed polymorphic blackening in industrial areas that condition is recessive in three instances only. One of these, affecting *Lycia hirtaria*, has just been mentioned, while the other two have both occurred in the Oak Eggar moth, *Lasiocampa quercus*. This is a large insect. The males are day-flying and about 6 cm across the expanded wings. The females are larger, about 8 cm across, and fly at dusk. The Oak Eggar is widespread in Britain, but in northern England and Scotland it is confined to moorland. It is abundant on Rumbold's Moor, Yorkshire, where it has been the subject of close study. The locality is heavily polluted by soot from Leeds and Bradford, 23 km away, and the moths there are greatly preyed upon by gulls.

The Oak Eggar males are normally reddish brown, the females paler; but about 4·7 per cent of both sexes are now of the greenish-black form *olivacea*, owing to a gene recessive in effect recently spreading in that locality. The females, in particular, gain protection from bird predation in their dark form as they rest upon the begrimed heather.

It is now necessary to mention an additional way in which moths sometimes respond to the conditions of manufacturing areas. That is to say, the caterpillars as well as the perfect insects may become blackish there. This is happening on Rumbold's Moor, where about 2 per cent of the Oak Eggar caterpillars are blackish instead of their normal foxy brown. Their dark colouring is also a simple recessive.

The blackening of these two stages in the life history of this moth are not aspects of the same condition. They are due to distinct genes: normal caterpillars can produce black moths and normal moths can arise from black caterpillars. If the two genes were un-associated, one would therefore expect that only 2 per cent of the black moths would be the product of black caterpillars, though in fact the value is about 50 per cent. It seems then that the two distinct genes are being built into a supergene (pp. 82–3), as is to be anticipated in polymorphism. We shall on p. 151 encounter an enticing possibility of detecting the same process in man, and we shall find also how important is the supergene in human genetics.

Pollution of the Soil by Heavy Metals

Two aspects of this need to be considered. In one of them, the contaminating material is airborne, a situation appropriately handled first, since it provides another aspect of what has been discussed so far in this chapter.

The pollution of the soil by metallic particles carried in the air has been studied by Lin Wu and others (1975). Since heavy metals can be extremely toxic to vegetation even at low concentrations, copper especially so, plants can provide the first indication of a serious situation of this kind. As these materials accumulate in the neighbourhood of metal refineries, species growing nearby are eliminated in a regular order; leading to the situation in which lawns and meadows are composed only of the grasses *Agrostis tenuis* and *A. stolonifera*, which can achieve a high resistance. Finally even these may fail to survive. There is then a total destruction of plant life; as at some places near a metal refinery at Prescot, in south-west Lancashire where, as a result of aerial distribution, the content of copper in the soil has now reached 4,000 parts per million.

The result of such contamination depends of course on a number of features: its intensity and nature (copper is more destructive than zinc); the length of time that the area has been exposed to it; and the ability of local plants to build up genetic resistance. As will be indicated in regard to the spoil tips of mines, some species seem incapable of doing so while others can achieve this to varying degrees. Thus the type of genetic variation available among the flora, as well as the force of selection, is of importance in places where the soil is being poisoned in this way.

Soil pollution by heavy metals may also be due to mining. All over the world this leads to the formation of spoil tips, unsightly and uselessly occupying perhaps valuable land. The genetic aspects of this matter have been examined by A. D. Bradshaw (1965) and his colleagues. They have worked principally in North Wales, but their results are of wide application.

Many ores of heavy metals, zinc, lead, copper and others, have been extensively mined in that district; the resulting refuse, poisoned by these substances, is deposited at the sites. They remain permanently unusable and largely uncolonized by plants: a blot on the country-side, looking like the landscape of the moon. Yet here and there in these derelict areas, Bradshaw noticed a plant or two had established itself. These he tested and found tolerant to the poison in the soil, their resistance being genetic and multifactorial, usually with a high degree of dominance. It is now established that since many genes are involved, slightly resistant forms exist at a low frequency, 1 per cent or so, in normal plant populations; and it is from these that effective tolerance can be built up.

In general, this must be done independently for each heavy metal; except for nickel and zinc, for which immunity to the one extends to the other. This is curious, for the two are not closely related chemically. Since the tolerant and non-tolerant plants take up the metallic compounds to an equal degree, any resistance to them must arise within the tissues.

Those capable of evolving such tolerance have no close botanical affinities: they occur in a wide range of higher plants, including grasses. On the other hand, some species, though growing up to the very edge of the mine sites, seem never to have produced a single specimen capable of colonizing them; of this the grass *Dactylis glomeratus* is an example.

The selection involved is powerful and accurate. A well known instance, studied in detail, is provided by the spoil tip of a small copper mine (300 × 100 m) at Drws y Coed, North Wales. It is situated in a valley which tends to canalize the prevailing west winds. Here the grass *Agrostis tenuis* passes from its normal susceptible condition to full tolerance in 1 m exactly at the western edge of the site. Downwind, tolerance fades out over 150 m of uncontaminated ground: evidence of windborne pollen; also of powerful selection against non-tolerant plants on the copper-charged waste material and, though to a less extreme degree, against the tolerant ones on ordinary soil.

Several species (e.g. the grass *Anthoxanthum odoratum*), having evolved tolerance to the poisoned conditions of mine tips, have there responded to selection for earlier flowering and increased self-fertilization. In that way, they have built up partial isolation from the non-tolerant plants in the immediate neighbourhood, so tending to prevent the formation of ill-adapted variants by crossing (see also p. 172). As might be expected, this effect is greatest on small mines and near the edge of large ones.

Bradshaw had originally considered the possibility of estimating the rate at which tolerance can evolve from the age of the mines at which it has become established. He has, however, doubtless correctly, decided that any such assessment might be unreliable, since it is possible that the soil at these places was already to some degree contaminated before the mining started. Fortunately some light on this matter has been obtained by other means. For instance, A. S. Watt erected an enclosure of galvanized wire netting at Lakenheath, Suffolk, in 1936. It was renewed in 1958. Subsequently, it was found that grasses growing at the foot of it were zinc-tolerant compared with those a metre or so away; an adaptation which had been achieved in less than thirty years. Bradshaw's laboratory studies suggest, moreover, that effective tolerance can be established in about five plant generations.

Adaptation of this kind has sometimes taken place in natural conditions. One finds, for instance, plants adapted to the high magnesium content of serpentine rocks; as at the tip of the Lizard peninsula in Cornwall, and on the serpentine outcrops here and there along the coast south of San Francisco.

It is reasonable to ask if genetic studies hold out any prospect of reclaiming land devastated by the waste from mines: a question of relevance in any part of the world. Bradshaw and his colleagues have accumulated large stocks of tolerant grasses and grass seeds. These grow well on the contaminated soil to which they are adjusted, provided they are supplied with abundant fertilizer. Here, immediately, one step can be taken by seeding: that of eliminating the blemish on the countryside which such minetips seem almost indefinitely to constitute: in addition some cover for animals would be provided. How far such plants would be available for grazing remains to be seen. It has already been mentioned that those which have gained immunity to the ores of heavy metals nevertheless take up these substances into their tissues. This may not be so for all soil poisons to which immunity can be gained, while it is possible that animals may successfully eliminate some of these materials when fodder containing them is eaten.

Water Pollution and Pest Destruction

Genetics could to some extent contribute to the important problem of water pollution, though it has not yet done so. It will nevertheless be useful to think briefly of the possibilities and limitations of such work.

It is likely that an extended study of blackening in moths may provide the earliest warning of air pollution in a number of its aspects. Not so genetics in respect of polluted water, which can so much more directly be traced to its cause: various effluents from a town, ill-controlled waste passed into a river from a factory; these are for chemists to detect, and for water boards and sanitary inspectors to take legal action to control. There is, however, the serious problem of the destruction of fish and water plants in lakes and rivers, even when pollution has been reduced as far as is reasonably practicable.

We do not possess extensive knowledge of fish genetics; but such information as we have shows instances of variation in a wide range of characters, including habit, multifactorially controlled, sometimes involving partial sterility; also polymorphism with heterozygous advantage. This has been found in a number of species, freshwater and marine, both by means of visible features and electrophoresis.

In view of these facts, but far more on general grounds, there can be no doubt that it would usually be possible to raise resistant stocks of many kinds of fish. Indeed in other forms we encounter that type of adaptation widely, because of the almost unlimited diversity of genetic variation.

For example, as soon as myxomatosis, introduced intentionally into Australia, accidentally into Britain, began to destroy dramatically the rabbit population, biologists predicted that the effect would not be permanent. Soon indeed animals arose that qualified by the development of a different physiology to withstand the disease, while simultaneously the infective organism produced less virulent strains. From then on, mortality among the affected animals was drastically reduced. In fact in South America, where myxomatosis is endemic, it is no more lethal to the indigenous rabbits than is chickenpox to British school children.

We find a corresponding situation when using insecticides against houseflies and mosquitoes. Resistance arises within two years and follows a well known course in two stages. First a resistant form due to the action of a major gene begins to spread rapidly. Later its action is enhanced owing to the selection of modifiers of the gene-complex. It is desirable therefore always to employ two such substances at the same time, say DDT and malathion, and to keep another in reserve.

The use of warfarin to kill rodents is also relevant here. This is an anti-coagulant, similar in effect to vitamin K deficiency, for it interferes with the action of that substance. We might therefore expect that genetic resistance to it is fairly widespread in mammals, including man; and so it has proved. The common rat, *Rattus norvegicus*, gains immunity to it from the action of a major gene: not so the house mouse, *Mus musculus*, in which the immune response is multifactorial and therefore takes place more easily. For there will always be a certain number of slightly resistant individuals available, from which full resistance can be built up. The mind returns to the occurrence among normal plants of a few specimens slightly adjusted to growing on soil contaminated by ores of heavy metals (p. 103).

So among fish; there will always be occasional specimens capable, to a greater or lesser degree, of withstanding the particular poisons that have depopulated rivers, unless their action be fully lethal at concentrations that cannot be reduced. If not, fish with improved resistance could be produced by selection and multiplied so as to restock relatively polluted water.

Of course the problem has wide ramifications. On what do the fish feed? It may be necessary to produce resistant plants: also Crustacea, such as *Gammarus pulex*. But here lies a remedy to one aspect of water pollution. Will it be thought worth while to apply it?

One general feature is to be noticed as organisms, whether plants or animals, become adapted to survive in polluted conditions of any kind: that is to say, the remarkable speed of the process. There are several reasons for this. One, of course, is the profound effect of the changes involved, often more considerable than those encountered in passing from one part of the world to another. In addition, there are special and favourable attributes of genetic variability, whether multifactorial or polymorphic: the one allowing a build-up from genes slightly appropriate to the new conditions (pp. 105–6), the other ensuring that when a major gene becomes an asset it will quickly reach a considerable frequency at which selection can act powerfully upon it.

Certain other components of genetic variation are not available in this connection: in particular, those contributed by mutation itself, as well as by the segregation of rare genes disadvantageous both as heterozygotes and homozygotes. Though individually very uncommon, these latter may be considerable in the mass.

On the other hand, what may, somewhat dangerously, be called 'pre-adaptation' cannot always be ignored. It may be that a variant originally favourable in one environment can, in the end, fit another. Thus Kettlewell points out that some black forms of moths could have had a selective advantage in the dark conditions of the pine woods that at first constituted the forests following the last Ice Age. These melanics may yet linger in the remnants of such habitats that survive today; for instance, the Black Wood of Rannoch in Perthshire. As he stresses, such anciently successful variants (e.g. the black form of the Mottled Beauty moth, *Cleora repandata*) may now acquire a new function among the horrors of urban surroundings and industrialism.

It should now be evident that the genetic adjustments forced upon organisms in polluted country throw light upon the process of evolution as it takes place at the present time.

Finally, at the end of two chapters dealing especially with polymorphism a simple method of detecting heterozygous advantage may be mentioned. The proportions of the three genotypes of autosomal alleles can be estimated in any population assuming they all survive equally well. For then, one homozygote, the heterozygotes and the other homozygote will be distributed as $p^2 : 2pq : q^2$. For example, if 1 percent. of a population be recessives, the percentage values of the three groups will be 81 : 18 : 1. An adequate sample will therefore enable us to say whether or not heterozygous advantage is operating, since we can compare the actual proportion of heterozygotes with that expected if they survive no better than the homozygotes (p. 153). See p. 46 for the comparison between *sex-linked* recessives in the two sexes.

6. Genetics in the Countryside and Garden

Wild Plants and Animals

The principles of genetics underlie both botany and zoology, and link the two subjects together. Yet in this matter there are exceptions, and among them may especially be mentioned *polyploidy* and *incompatibility*: widespread and important among plants, rare or absent among animals. Indeed so well known are they that gardeners and those who grow fruit or vegetables commercially may often refer to them in casual conversation, so mystifying the animal breeder or the zoologist.

It will be convenient to describe polyploidy at the outset of this chapter, leading on to a brief survey of genetics in the countryside. A short account of incompatibility can then be used to extend that picture and afterwards to introduce the genetics of cultivated plants.

Variation in Chromosome Number and Origin

The chromosomes, and therefore the genes they carry, are normally present in pairs; two of each type, derived respectively from the two parents. This is the *diploid* situation (p. 25), represented as 2n. It arises by fertilization in which the two gametes, being the male and female reproductive cells of plants and animals, fuse. For these normally carry a single member of each form of chromosome (p. 25); the so-called *haploid* number (n). When, however, the chromosome content of the cells is multiplied to higher values, the result is known as a *polyploid*. It can be specified as a triploid (3n), a tetraploid (4n) and so forth.

The ordinary Mendelian system is based on the diploid arrangement. It is, of course, one in which the pairs of corresponding genes (alleles) are present in similar or dissimilar phases; that is to say, as homozygotes or heterozygotes. Evidently this will be modified in polyploidy, in which the gametes carry more than one member of each allele and the body cells more than two members; for then mitosis still tends to maintain the chromosome numbers and meiosis to halve them.

The genetic effects of polyploidy can be complicated. This is true even in the form easiest to understand, that of tetraploidy. Those interested in it can obtain a full account of the subject from Darlington and Mather's work and from Darlington (1963); also from Stebbins (1963). A description of polyploid segregation would be unsuited to a book such as this, which is intended for general readers. Indeed, to understand this chapter it is really necessary only to bear in mind two basic facts about polyploidy.

First, polyploids tend to be less fertile than diploids. This is due to the complex interlacing that can arise from their chiasma formation, and prevents simple segregation. Also, odd number polyploids (3n, 5n, and so on) are sterile to an extreme degree as a third allele per locus prevents uniform pairing.

Secondly, the polyploid situation gives rise to more complex results than the diploid one. That is to say, it produces more classes of genotypes in segregation and ensures that they appear in higher-value ratios.

This may be illustrated by an example showing its results, yet without dealing with the somewhat involved means by which they are brought about. Here, in the sixth chapter of this book, it will doubtless be realized that in ordinary diploids a mating between two heterozygotes (*Aa* × *Aa*) leads to the formation of three genotypes among the offspring (*AA, Aa, aa*) in a ratio of 1 : 2 : 1; while dominance converts this into two distinguishable classes in a ratio of 3 : 1. The comparable tetraploid situation is represented by the mating *AAaa* × *AAaa*. It leads to the formation of five genotypes among the offspring (*AAAA, AAAa, AAaa, Aaaa, aaaa*) in a ratio of 1 : 8 : 18 : 8 : 1 (= 36); while full dominance converts this into two distinguishable classes, not in a 3 : 1 ratio but in a ratio of 35 : 1.*

Yet it sometimes happens that though the effect of a gene is fully dominant in a diploid (*Aa*) and in the same proportion (*AAaa*) in a tetraploid, one dominant cannot entirely mask the effect of three recessives (*Aaaa*). For instance, in the primrose species *Primula sinensis*, white flowers are in the diploid fully dominant to red; but they are of an intermediate pink shade when one dominant is opposed to three of the recessives.

It will be noticed that a study of tetraploidy also involves gamete

---

* It does so because we are dealing with a ratio based upon 36 items of which the last, *aaaa*, is the only one without *A* (the gene producing the dominant character).

formation by the other two types of heterozygotes (*AAAa* and *Aaaa*) in addition to the situation here considered. Moreover, the arrangement may be much modified by crossing-over which, with the eight chromatids of a tetraploid, can produce a complex interlacing. Furthermore, when we take into account other polyploid values, 3n and those above 4n,* it is obvious that the matter is intricate. We must therefore restrict ourselves to the genetic consequences of polyploidy and omit its analysis. However, a few other basic points need to be established at the start.

First, it ought to be said that the step downwards to a haploid individual (n) cannot be taken along the lines of polyploid formation, but only by 'virgin birth' (parthenogenesis). Very rarely, a haploid gamete develops into an adult plant. Sometimes it does so as a result of imperfect fertilization; the nuclei fail to combine, though penetration by the male gamete stimulates the female one to grow. Haploid higher plants are very rare, weak and infertile. It is just possible that haploids may now and then be formed in higher animals. Were this to take place in a woman it would certainly not be correctly diagnosed, but would be characterised as an event, unfortunately, of no unfamiliar kind. There is a curious psychological tendency to surround the birth of some great men with the degree of mystery which virgin birth is held to supply, and it has even been suggested that parthenogenesis provides a 'natural' explanation for such an event. It is perhaps worth pointing out that in human reproduction it could only give rise to a female child (p. 40).

There are a few animals adjusted to develop as haploids. In the hive bee the worker caste is diploid and female like the queen, but sterile. The distinction between the two being produced by the food supplied to the larvae. The male, however, is derived from a haploid gamete. Its germ cells, and those of its nervous system, remain haploid, but the chromosomes in the other body cells increase not only to the diploid value but some even to 3n and 4n.

Polyploids are of two main types. The one to think of first, and indeed that so far described, is *autopolyploidy*. This can occasionally arise in a fertile diploid owing to a failure in meiosis. Being highly exceptional, the 2n gametes that it must form will almost always meet ordinary haploid ones (n) at fertilization, so producing triploid plants (3n); as of course will a cross between a diploid and a tetraploid. On the other hand, an error in mitosis in which the

---

* The series cannot be increased indefinitely: 6n is found, but though 8n can be produced and may result in a successful cultivated plant, it is not often effective in a wild one subject to competition.

nuclei fail to separate can establish a tetraploid body cell from which a bud may grow. Such 'bud sports' will bear tetraploid flowers and diploid gametes, so originating a line of tetraploid plants by selfing, if that process is possible in the species concerned.

Tetraploids can also be produced artificially. Errors in cell division are more likely than elsewhere to occur in the rather irregular growth of callus which forms where plant tissue has been wounded. That fact was used to produce the first artificial tetraploids, by cutting back the stem of tomato plants. It was found that about 6 per cent of the shoots that sprouted from the cut surface were unusually large, and that these were tetraploids.

Further, a number of drugs can interfere with mitosis in such a way that the chromatids do not segregate from one another, so forming tetraploid cells. Colchicine is a particularly efficient agent in promoting such abnormal cell division when young seedlings are placed in a dilute solution of that substance.

How do full polyploids establish themselves? There are difficulties here, for the initial chromosome abnormality will arise in a single plant only, so that, apart from selfing, its diploid gametes (2n) will cross with the normal haploid ones, (n), so originating *triploids*. These, though indeed often larger than their parents, are highly infertile because they have an extra chromosome in addition to each chromosome pair. As a result of chiasma formation, the gametes contain all combinations of the odd members, so giving rise to genetically unbalanced types. Yet triploid survival may to some extent be ensured by vegetative propagation, a tendency that will be favoured if the polyploidy has something of an advantage. So much is this true that at one time it was supposed that chromosome multiplication itself tends to induce vegetative growth, though this is not so. However, that process, purely selective as it is, does ensure that large stands of some polyploid plants can occasionally become established.

There are many nearly related species whose chromosomes are exact multiples of each other. As examples, one may cite *Gentiana nivalis*, the Small Gentian (2n = 14), and *G. verna*, the Spring Gentian (2n = 28); or *Lotus tenuis*, the Slender Bird's-foot (2n = 12), and *L. corniculatus*, Bird's-foot trefoil (2n = 24). This is probably the result of simple polyploidy; but here, and in other instances, it is not always clear how the condition has originated.

Another important point must now be raised. Those errors in cell division which multiply the whole chromosome set may affect one or more of the chromosome pairs only, so originating *polysomics*. These are indeed partial polyploids; triploids, perhaps, for a single chromosome type. In general, their genic balance is upset, and they die. However, it seems that in some species certain chromosomes can be present three times without results that are too serious. The plants that carry them may then persist long enough for two of these accessory members to be brought together by crossing. In that way, an extra chromosome pair may become established. For example, there are two European, and British, species of Sea Holly: *Eryngium campestre* (2n = 14) and *E. maritimum* (2n = 16). Great numbers of such instances are known in closely related forms. Also *monosomics*, that is diploids in which a chromosome is represented once instead of twice, can sometimes survive if it be small. If large, the lack of balance which results is too great, and kills them.

Since something like one-third of all flowering plants are polyploids or polysomics, it is evident that these states can be advantageous. What is the nature of these advantages?

Leaving aside for the moment the special qualities of hybrids (*see below*), polyploids tend to be larger and to produce larger flowers and fruits than diploids. This seems a direct consequence of their necessarily increased nuclear volume to accommodate the excess chromosomes; and since there is a relation between nuclear size and cell size, so also the plant itself is enlarged. These features can well be of advantage in competition. In addition, polyploids may contain excess quantities of certain valuable substances, such as vitamins and some types of proteins (Stebbins, 1963, p. 304). Moreover, as tetraploids can form a satisfactory colony or population among themselves, but not in crosses with diploids, even-number autopolyploidy can act as an isolating mechanism (pp. 74–5, 120). Against these assets must be set slower growth and reduced fertility. However, the reproductive potential of most plants so far exceeds what is generally necessary that its reduction may not be harmful.

The second type of polyploidy comprises the *allopolyploids*. These have the double chromosome set of a more or less sterile hybrid. The offspring of a cross between distinct but related species possessing, respectively, such a chromosome complement as AABBCCDD and EEFFGGHH are sterile because they carry one member only from each parent: ABCD and EFGH. Consequently these have no homologues with which to associate at meiosis. But by doubling the total as in an allotetraploid, to produce AABBCCDDEEFFGGHH, each chromosome gains at least a potential partner and germ cell formation can go forward. Ideally, such an individual is true-breeding, also self-fertile if that condition be appropriate to it; yet it is sterile with both its parental types: it has become a new species at a single bound.

This represents the basic concept of the matter, but actually it is

subject to a number of complications. These depend upon the extent
to which the chromosomes do in fact pair, and on the results of that
process (of this, Darlington and Mather, 1949, give a clear account).

In the true-breeding allotetraploid which has occurred in the
sterile hybrid between the radish and the cabbage, there is little
chromosome pairing. Thus it successfully combines, with no marked
deviation, some of the qualities possessed by its two parents. Now
consider the distinct *Primula verticillata* from Abyssinia and *P. flori-
bunda* from Afghanistan. In both, 2n = 18, and about two chiasmata
form between each pair of homologous chromosomes. The species
can be successfully crossed, giving a fairly intermediate hybrid in
which 2n still equals 18; in it chiasma formation is by no means
absent, though it is reduced to about 1·4 per pair. Meiosis therefore
is fairly normal, yet sterility is total. This, as Darlington and Mather
point out, is due to the fact that the situation is unbalanced because
some chromosomes are from one parent and some from the other.
Such hybrids have occasionally produced tetraploid shoots; they
did so first at Kew in 1899, to the interest of the old Duke of Cambridge. In these, 2n = 36, and they are fertile and true-breeding.
They constitute a new species, *Primula kewensis*; and this is sterile
with the parental ones. Its fertility, however, is not quite up to the
standard of the radish-cabbage allotetraploid; nor are the plants
quite so intermediate as the undoubled hybrid from which they
arise. For in the tetraploid, the chromosomes from the one species
generally pair together, as they should, rather than with those from
the other; yet the latter occurrence does also take place, producing
genetically unbalanced progeny which die, so reducing fertility.

It will be apparent, then, that allotetraploidy can lead to rather
diverse results. One of these must in particular be mentioned. This
is described as 'segmental allopolyploidy' by Stebbins. It is a condition in which two sets of chromosomes, from different species, have
nevertheless a considerable number of corresponding chromosome
segments, or even whole chromosomes; while, on the other hand,
they differ in a sufficient number of genes to be sterile as diploids.
They arise as ordinary (diploid) hybrids in which a certain amount
of chromosome pairing at meiosis does take place.

It should be added that autopolyploidy and polysomy can
obviously occur in an allotetraploid, just as if the latter were a
normal plant. The types can therefore integrate to varying degrees.

The question now arises of what advantages the allopolyploid
possesses. The answer is that, as already mentioned (p. 61), genes
interact with one another to produce the features for which they
are responsible. Consequently, when those which normally never
meet are brought together by exceptional means, they can evoke

new aspects of variation which may sometimes be useful, especially in a novel environment. It is for this reason, indeed, that a hybrid may possess characters absent from both its parental forms. Allopolyploidy may, therefore, do more than combine in one individual qualities derived from two distinct species. Yet even that may be economically desirable. We can easily see how valuable would be a plant having the leaves of a cabbage and the root of a radish. Unfortunately that particular cross has so far resulted only in bringing together the two reverse, and useless, types.

Allopolyploidy is more important in agriculture than autopolyploidy. It has often been pointed out that though a number of the crop plants may seem to be autopolyploids, the evidence suggests that they are as likely to be typical or segmental allopolyploids. Among these may be mentioned coffee, banana and sweet potato (*Ipomoea batatas*).

Colchicine (p. 111) can be used greatly to increase the production of tetraploid cells in a hybrid, just as it can in an ordinary plant. It thus becomes possible to exploit the advantages of allotetraploidy more fully; a technique likely to be of commercial value. It must be recognized, however, that a new species, whether stimulated to appear in this way or arising through a spontaneous chromosome abnormality, will at first be ill adjusted, and will require a period of selection to fit it for commercial purposes before it is likely to be successful, whether in horticulture or as a crop plant.

Considering polyploidy in general terms, we encounter the remarkable fact that though the condition is extremely common in plants, it is rare in animals (p. 108). In a first attempt to explain that anomaly, it was suggested that the polyploid must break down the usual animal sex chromosome mechanism, which is a balanced one. This may be correct in certain instances, but it does not solve the problem as a whole. For instance, that particular difficulty does not apply to hermaphrodite animals, yet polyploidy is rare in them also. A sounder view suggests that the sudden alteration of the genotype that polyploidy involves must produce disharmony in the processes of animal development, which are so much more complex than those of plants.

A further point is valid here. Tetraploidy arising as a bud sport from a cell with a doubled chromosome set is excluded in animals. In them, it must be attained by the alternative mechanism: the formation of an unreduced (2n) gamete. The product, being a rarity, will meet the normal reduced type, resulting in a generally infertile triploid. Yet sterile triploids in plants can be maintained by means of vegetative reproduction, a situation virtually closed to animals.

Incompatibility

Close inbreeding, the closest of all being self-fertilization, leads towards genetic uniformity. On the other hand, outbreeding, involving wide crosses between different individuals, results in genetic diversity. In what circumstances are these two conditions, or intermediates between them, respectively desirable? Evolution is possible only if there be genetic variation upon which selection can act. For such variation is at random relative to the needs of the organism, and random changes can rarely indeed do other than damage a highly adjusted system such as the organization of a plant or animal. Without selection there would be destruction, for neither recombination or mutation is directed to what the body requires.

On the other hand, if an organism is well adapted to its present habitat, it must from the short-term point of view be advantageous for it to preserve that situation and remain constant; for even selection is then to an extreme degree unlikely to improve what is already satisfactory. Inbreeding, therefore, should be encouraged in such circumstances. Yet in the end it is fatal; for the environment is bound to change, and an organism deficient in variability cannot adapt to new situations and must perish. It cannot survive unless selection is removing the less well adapted forms and adjusting in the right directions those that are more favourable. Some balance should therefore be struck between different breeding systems. Indeed, the perfect one would be that in which a species can pass as required from outbreeding, tending to diversity, to inbreeding, tending to uniformity; or the reverse. We actually encounter that auspicious type of adaptation (pp. 117–20).

Sexual dimorphism, in which the male and female gametes are formed by different individuals, encourages outbreeding with diversity. This, though found in a great diversity of plants, is nevertheless rare (it occurs in about 2 per cent of flowering plants). Yet it is the general situation in animals. Why should this be? Mather has suggested that wastage is reduced when mate-discrimination is effective, owing to animal mobility.

There is, however, an efficient means of securing outcrossing in hermaphrodite plants: that is to say, by means of self-sterility, in which no seed is set when self-pollination occurs. The system can also be extended in such a way that crosses are successful only when they take place between plants of certain genetic types, a situation known as incompatibility.

This is achieved by controlling the rate at which the pollen grows

down to the ovary of the flower from the stigma* where it germinates. Normally pollen cannot grow, or grows too slowly, in a flower
carrying the same incompatibility genes as itself. These genes,
known as S, comprise a multiple allelic series which, astonishingly
enough, sometimes includes more than 30 alleles.

They are supergenic, of two parts determining respectively male
and female incompatibility. If both the male and female plants
possess, for instance, $S1$ and $S2$ (one carried by the pollen, the other
possessed by the stigma) fertilization fails; if one sex possesses $S1$
and $S2$ and the other $S3$ and $S4$, compatibility is complete. If, on
the other hand, one sex has $S1$ and $S2$ and the other $S2$ and $S3$,
half the cross is compatible ($S1$ and $S3$) and half ($S2$ and $S2$) is not.
For instance, in cherries, Early Rivers is incompatible and therefore
sterile when pollinated by itself or by Bedford Prolific, but compatible and fertile when pollinated by Black Heart. To ensure a
crop, it is necessary to combine the correct varieties in proximity
when planting an orchard. The usual horticultural books, and
salesmen at plant nurseries, provide information on which compatibles can appropriately be grown together.

The incompatibility mechanism so far mentioned presents one
side of the picture, favouring outbreeding with diversity. But what
about the other aspect, leading to inbreeding with uniformity?
This, too, is catered for by means of a fertility gene, or genes (S^f),
ensuring dominant self-compatibility. It may be a member of the
multiple allelic series already mentioned. In those circumstances,
$S^f S^x$ (where x is any other member of the series) is effectively self-
fertile. On the other hand, the fertility gene may be at another locus.

We see the need for, and the working of, this self-fertility mechanism when, for example, plants brought to a new habitat are removed
from the insects that pollinate them. Thus, the tomato crosses
freely in wild conditions in South America, but is normally self-
fertilized in England. An additional insurance in this direction is
secured when the buds of a plant fail to open, or do so only after
the pollen is shed: as in the sweet pea in northern Europe, but not
when wild in Sicily (p. 113).

This is not a book on fruit growing, but a few remarks of a more
practical nature may be made on incompatibility. The reaction is
a reciprocal one. When we cross two diploids, one with $S1S3$ and
the other with $S1S4$, it makes no difference which is carried by the
pollen and which by the plant receiving the pollen. A tetraploid
will, of course, have four incompatibility loci in its body cells and
two in its germ cells.

---

* A female part of the flower on which the pollen is received.

Compatibility and incompatibility genes are widespread in fruit trees and among plants generally. Most fruit trees are pollinated by insects, in Europe especially by bees. There is, of course, no barrier to growing together large blocks of compatibles, such as the Victoria plum; but when using self-incompatibles, large blocks are to be avoided and interplanting with carefully chosen types is necessary. The choice of apple trees requires particular care, since they are remarkable for an almost continuous range of incompatibility. It is to be noticed that the pollen of self-compatible forms appears to be effective on incompatible ones. Incompatibility genes operate in a wide range of plants, and are of importance not only to the horticulturalist but to the farmer; as in clover.

Genetics in the Countryside

It would be impossible in the space allowable here to give even the briefest account of so vast a subject as genetics in wild plants and animals. But some of the main types of genetic variation that have up to now been studied in wild organisms may usefully be illustrated at this point by one or two examples of each. They will suggest what may be expected on, as it were, taking genetics out into the field. But first we can trace incompatibility, and then polyploidy, a little further in wild forms than can conveniently be done experimentally or under cultivation.

In the incompatibility so far described, the different mating groups are alike in structure and appearance: it is purely the breeding reactions of the individual plants that are affected. This surely is the fundamental arrangement, but in many species there is added to it an anatomical distinction that increases its efficiency because it reduces wastage. It is one in which, though the plants are hermaphrodite, there is a barrier to the deposition of pollen on the stigma of the flower that has formed it. But the mechanism has a wider application: it largely checks the, necessarily unsuccessful, pollination of incompatible flowers on other plants of the same species.

This arrangement was originally studied in the Cowslip, *Primula veris*, and the Primrose, *P. vulgaris*, by Darwin. We may concentrate upon the latter. Here we find two types of flowers: 'thrum-eyed' and 'pin-eyed', and each plant produces one or the other but not both.

On looking into a primrose blossom, one sees in the centre of the petals a small, round hole which may be largely filled by something rather like the head of a pin. This is the stigma carried at the end of the style: a 'rod' that leads down to the ovary at the bottom of the tube-like base of the flower (technically the 'corolla tube'). This is a 'pin flower'. Alternatively, the place of the pin may be taken by five small, pointed structures arranged in a circle. They are the anthers in which the pollen is formed: this is a 'thrum flower'. The distinction between such pin and thrum primroses and cowslips is well known to country children.

These two types of plants are found in something like equality, but with a small but definite excess of pins. A random count carried out by Mr. and Mrs. Beaufoy in Suffolk amounted to 1,827 pins and 1,553 thrums.

If we tear open a pin flower down its length, we find a long style arising from the ovules and bringing its enlarged distal end, the stigma, level with the top of the corolla tube, halfway down which are the five anthers, corresponding with the five petals. On examining a thrum flower, the positions of anthers and stigma are seen to be reversed. The latter is halfway down the tube because the style is short, and the anthers are at the top.

Now Darwin pointed out that this arrangement tends to encourage out-crossing. The primrose, in common with other flowers of this type, is known as 'distylic' because it has these two floral forms. They are pollinated by insects with a long proboscis: moths, *Bombylius* (a species of fly as large as a bumble bee and with rapid movements) and some others. The pollen from the thrum anthers sticks round the base of the proboscis when this is pushed down the corolla tube to reach the nectaries at the bottom, and may subsequently be deposited on the high stigma of a pin flower, with its low anthers. From the latter, pollen grains can be picked up near the tip of the proboscis of an insect visitor, to be transferred later to the low stigma of a thrum flower.

The plan certainly tends towards out-crossing, but it is only moderately successful. The withdrawal of the proboscis from a pin flower may carry pollen up to its high stigma; while pollen from the high anthers of a thrum flower may easily be carried or fall down to the stigma below. Moreover, insects with a long proboscis are not the only pollinators to be reckoned with: small beetles often inhabit the flowers, and scatter the pollen indiscriminately.

The features involved in this 'heterostyle'* mechanism are not limited to the positions of the anthers and stigma. For instance, pin pollen grains are only two-thirds the diameter of the thrum type. Consequently, examination with a lens will show to what extent the two forms are mixed. However, the most important

---

* So called because the style is of two types, either short or long.

feature of the heterostyle system is the extension into it of *incompatibility* similar to the kind, already discussed, which is found in flowers that do not respond to it by a difference in structure.

If thrum pollen be placed on a thrum stigma, it is almost completely infertile. The same is true, though to a lesser degree, of pin on pin. This barrier to fertility is of two kinds. Though thrum pollen germinates well on a thrum stigma, it usually fails to penetrate its surface. Pin both germinates on a pin stigma and penetrates it. But the like (pin) pollen grows down to the ovules at the bottom more slowly than the unlike (thrum) pollen. Consequently the latter reaches the ovules and brings about fertilization first when the two kinds are mixed, as they will often be in nature. That is to say, crossing is successful when pollen is from anthers at the same level as the stigma, but not when they are at different levels, as they will normally be in the same flower; therefore, the system leads to outbreeding with diversity.

Heterostyly is controlled by a supergene whose components are so closely linked that the system usually behaves as if it were due to a single unit. S gives rise to the thrum type, which is dominant, and s to the pin. Since heterostyly ensures that, in normal circumstances, crossing must be between unlike types, it takes the backcross form $Ss \times ss$. Thus, thrums are heterozygous and pins homozygous. The two are produced in approximate equality, and the homozygous thrum (SS) does not occur.

We can now look at the other side of the picture, that in which this system can be used to produce inbreeding tending to uniformity.

The supergene S and s is built up of a number of closely linked loci. Rare crossing-over can occur within it, taking place between the blocks of loci controlling the male and female parts of the flower, *with those for their incompatibility*. These may be represented as M for the dominant male type* in thrum, and m for the recessive male type in pin; also F for the dominant female type† of thrum, and f for the recessive one of pin. Thus the supergene for pin, with its structure and incompatibility, may be thought of as $ss = (mf)(mf)$; and that for thrum as $Ss = (MF)(mf)$. Crossing-over within the latter results in $(Mf)(mF)$. This produces gametes for short 'homostyles' (Mf) with anthers and stigma low, and long 'homostyles' (mF) with anthers and stigma high. Since the incompatibility genes are transferred with those for the sexual structure, such flowers

---

* That is, anther position = male type; stigma position = female type.

† So called because the style is of one type, with anthers and stigma at the same level.

become self-fertile and self-pollinating. The mechanism, ensuring relative variability and the alternative of relative stability, seems almost miraculous in its perfection.

Primroses are heterostyled, of the type which produces out-breeding with variability, throughout Britain (so far as is known) save in two small areas where a high proportion is homostyled. These sites are about 130 km apart: one of about 26 by 23 km in Somerset; the other a little smaller in Buckinghamshire. In both, it is the long homostyled form that has established itself. The few heterostyles remaining in these special areas are chiefly pins. For from what has been said of the behaviour of incompatible pollen in the primrose, it will be evident that pins could remain in the absence of thrums but not the reverse (p. 119).

We do not know what aspects of the Somerset and Buckingham-shire localities just mentioned favour stability rather than variability. It can only be said that the whole of the English countryside has departed so far from its primaeval state, and has been so greatly altered by changing agricultural practice within the last two centuries, that it is not surprising to find some districts where the primrose is so well adapted to its immediate environment that it is better for it to remain relatively invariable.

The heterostyle-homostyle system occurs in many plants. Its value can be seen also in another context in the genus *Primula*. This contains five species in Britain. Four are normally heterostyled, while the little Scotch primrose, *P. scotica* (6n = 54), of the far north of Scotland and Orkney, is a completely homostyled species. It is obviously derived from the Bird's-eye primrose, *P. farinosa* (2n = 18), which it greatly resembles. The latter grows in damp places on the mountains of northern England and Scotland.

It would seem that the homostyle system has enabled *P. scotica* to preserve its adaptations to the far north. To that end, its poly-ploidy will have contributed. Indeed it may be that its capacity for selfing has made its hexaploid condition possible.

There is a curiously complex situation, the 'tristylic' one, in which three floral types are involved in incompatibility. It is found in the Loosestrife (*Lythrum*), Daffodils (*Narcissus*) and some other groups. Here the sexual organs grow at three lengths within the flower: that is to say, the style can be long, 'mid' (intermediate) or short, while the anthers can occupy three corresponding positions. The latter can take any two of these sites, with the stigma at the other one. Since all the flowers on a plant are of the same kind and since, as with the primrose, fertile unions can be obtained only when pollen is formed at the same position as the style, the plan ensures outbreeding.

It is controlled by two pairs of unlinked genes. One determines whether or not the style shall be short. The other has no effect upon short styles, but decides whether the non-shorts shall be mid or long. The latter is the double recessive and the short is the double dominant. Therefore mid styles are dominant to long but recessive to short. The system is adjusted to different proportions from one wild population to another. It is sometimes combined with polyploidy, as in the Purple Loosestrife, *Lythrum salicaria*. It is not clear what are its advantages compared with the distylic arrangement, except that it increases heterozygosity.

Polyploidy presents a practical problem to the researcher: since we generally see this when long established in wild populations, it is not often that it has so far been possible to study its evolution in them along lines, in fact, somewhat similar to the process that produced *Primula kewensis* in experimental conditions (p. 113). Consequently, a fine example of the kind is well worth mentioning; especially as it has economic effects of some importance, while it may provide a pattern for what awaits discovery in the countryside today.

The Cord Grass *Spartina* is a plant adapted to growing on tidal mud flats. The British species with which we are concerned is *S. maritima*, with 60 chromosomes. At some time about 1870 it was found to have crossed with the American *S. alterniflora*, the chromosome number of which is 62, one pair being represented twice. This was first reported in Britain about 1840 near Southampton, where it had been accidentally introduced, probably in the ballast of ships. The two species produced a sterile hybrid, reproducing vegetatively, with 62 chromosomes; 61 would have been expected, but it is likely that the chromatids formed by the extra pair had failed to separate at meiosis. Chromosome doubling to 124, in some instances to 120, restored fertility. Subsequent back-crosses seem to have produced a complement of 92 (as 62 + 30). From this, a further back-cross has given rise to plants with $2n = 76$ $(46 + 30)$. These behave as a new fertile species known as *S. townsendii*, sterile both with *maritima* and *alterniflora*. It has qualities absent from each of these (pp. 113–14) which have given it superiority over them. Consequently, it has flourished to an extent which is causing it to be a serious pest today, choking, for instance, Poole Harbour.

Since differing chromosome numbers promote isolation, there should be, and perhaps is, a tendency for polyploids to occur where distinct races or closely related species come into contact (pp. 112, 120). It would be valuable to look into this further than has so far been done; including, especially, those instances where interspecific hybrids are known to have formed, as in the genus *Erica*, the heaths. The relative distribution of polyploidy is discussed by Stebbins (1963, pp. 342–50). He finds, among other things, that it tends to be particularly common in plants occupying areas newly opened up to them. This, on the whole, confirms what has just been said; for it will be important to secure adaptation to such habitats by means of selection, a process that generally requires isolation.

It has been said that polyploidy is largely restricted to plants (p. 114). Yet a search for it among wild animals does reveal a few instances of the condition.

The chromosomes of the three European Wood White butterflies have been studied. In *Leptidea sinapis*, the range of which extends to southern England and Ireland, the diploid number is given as 26, occasionally 41;* while in *L. duponcheli* 2n = 104. The three European members of the genus, with one in Asia, form a mysterious little Old World group, since they belong to the Dismorphiinae, a subfamily of close on one hundred species otherwise restricted to South and Central America.

Another example is found among the Lycaenid ('Blue') butterflies: *Lycaena bellargus*, the Azure Blue, in which 2n = 45; and *L. coridon*, the Chalk-hill Blue, with a diploid number of 90. These two are sufficiently close for occasional specimens which appear to be natural hybrids between them to be reported. Their isolation is achieved at least in part by their times of appearance on the wing, the single brood of *coridon* falling between the two broods of *bellargus*. Many other sporadic instances of the kind might be quoted. However, as far as is known, there is but a single animal group, the Salmonid fishes, in which there is a complex mass of polyploids comparable to that in certain plants (e.g. the Blackberries). The reason for this is not yet clear.

On the other hand, evolution based on polysomic variation is not uncommon among animals in natural conditions, especially when belonging to races or species that are restricted in their habitats. It can be highly suggestive, as the following example will show.

In Palestine, the Mole rat, *Spalax ehrenbergi*, occurs in four chromosome races identical as to the appearance of the animal, but apparently related to increasingly arid conditions (Wahrman *et al.*, 1969). From Mount Hermon to the Sea of Galilee and west to the Mediterranean 2n = 52. At the same latitude but east of the Jordan 2n = 54; from the Sea of Galilee southwards to Samaria 2n = 58, while further south still, as far as the Negev, 2n = 60. Though the races adjoin, very few hybrids have been reported; indeed the

* It is possible that two species similar in appearance are concealed here.

subterranean habits of this creature probably favour the evolution of local adaptations.

As I have pointed out elsewhere (Ford, 1976), if the establishment of these chromosome types could be studied still further, it might throw light upon human evolution. In Man $2n = 46$, while in the Apes most closely related to him, the Chimpanzee, the Gorilla and the Orang-utan, $2n = 48$. Furthermore, in the Gibbon, which is the next nearest to the human line but considerably further from it, $2n = 44$. Monosomics and polysomics, the building materials of such distinctions, though rare and unfavourable, are well known in man (p. 163). There is a sufficiently obvious comparison here to make the cytology of *Spalax* of interest in the history of our own evolution.

In considering the main types of variation that have been examined in wild plants and animals, it will be recognized that a number of these have already been discussed in this book and it would therefore be inappropriate to introduce them again. Those that may particularly come to the mind relate to industrial melanism (pp. 92–102), plants growing upon the spoil tips of mines (pp. 103–5) and the genes responsible for mimicry (pp. 87–9). Indeed, many other examples of genes in wild populations have been mentioned, especially in the chapters dealing with evolution. Here additional information on some of these types can be given and others added.

The multifactorial situation is of course widespread, allowing easy adjustment to diverse conditions, since it ensures that there are likely to be some appropriate genes present in the normal population of a species from which necessary adaptations can be built up. This can be illustrated from resistance to warfarin, used for destroying rodents. Such resistance is acquired much more quickly in the House Mouse, in which it is multifactorial, than in the Rat, *Rattus norvegicus*, in which it is controlled by a single gene.

The Sea Plantain, *Plantago maritima*, is remarkable for extreme variability, multifactorially controlled, which fits it to highly distinct environments: these range from waterlogged mud in salt marshes to crevices in dry rock, with situations of every intermediate kind. The adjustments are mainly physiological, but affect also such features as the length of the flowering stem and the shape of the leaves. They are 'continuous' in the sense that they can take a complete range from one extreme to the other. Discontinuous variation multifactorially controlled has, however, already been discussed in the spotting of the Meadow Brown butterfly; for this falls into the series from no spots to five spots.

It goes without saying that many of the rarer variants in plants and animals are due to segregation, giving rise to features that tend to be eliminated by selection. When recessive, such qualities appear sporadically, as in plants with coloured flowers in which white forms are found from time to time. One may mention also the rare yellow variety of the Green-veined White butterfly, which is due to a gene which removes one oxygen atom from the molecules responsible for the white colouring. Moreover, interactions may take place between these and other striking features. For instance, in the Bladder Campions, *Silene*, white-flowered variants normally behave as simple recessives, but when two are crossed they may have coloured flowers. Each P1 parent will then have contributed an essential gene for colour lacking in the other.

When we turn to geographical variation, the distinctions to be encountered from one region to another may have evolved in several ways. They may represent adjustment to different environments, or they may have originated independently; and it is sometimes the case that similar results have been reached by distinct means in one region and in another. Experimental analysis is generally required to separate the two types.

In England the males of the Muslin moth, *Cycnia mendica*, are blackish and the females white; while in Ireland, and in a few isolated localities in Continental Europe, the males are white like the females. The two forms are determined by a single pair of alleles, sex-controlled in effect and without dominance: the males from a cross between English and Irish parents are of a sandy colour. Evidently, the races have been perfected also by selection acting upon the gene-complex.

It has been pointed out that plants often become polyploids towards the edge of their range (p. 120). This ensures that the special adaptations required in such extreme conditions can evolve accurately in (genetic) isolation. One of the Valerians, *Valeriana officinalis*, adjusted to dry calcareous soils is a diploid (2n = 14) across the great plain of Europe. It becomes a polyploid towards the western edge of its habitat, and only tetraploids have reached England. Here they have produced octoploids which have the same chromosome number as a distinct species, *V. sambucifolia*; these can then cross. They produce a wide range of hybrid types, some of them even able to colonize wet and acid soils.

It is by no means unusual to find a species polymorphic in one part of its range but not in another, a situation that can be illustrated from the Scarlet Tiger moth, *Panaxia dominula*. This is a highly localized species forming isolated colonies, generally in marshes and along river banks. Flying during the day, and having scarlet hind wings and fore wings of a metallic blackish-green colour with white spots, it adds an almost tropical touch to the countryside even in England, where it occurs from the Midlands southwards.

The species is polymorphic for a gene affecting its pattern in a single marsh about 8 km from Oxford. There, in addition to the normal form, two variants are established. These are a rare homozygote, *bimacula*, and the visibly intermediate heterozygote *medionigra*. This colony has been studied quantitatively by my colleagues and myself for thirty-nine years. Throughout that time, the number flying each season has been calculated by the method of marking, release and recapture, while the percentage of the *medionigra* gene has also been recorded. The results have differed annually, there being one generation a year. Now with Mendelian inheritance, and knowing these facts about the population, it is possible to calculate how great a change in proportion could arise merely by chance survival from one generation to another. The observed fluctuations in fact considerably exceeded what could be produced in that way; they must therefore have been due to natural selection, varying in intensity and direction from year to year. This was the first time in which it had been possible to estimate the importance of selection compared with random survival in any wild population.

When plants or animals occupy a continuous range from one type of country to another, they may form a cline (p. 72), taking the form of a short or long sequence of intermediates, while its position may shift in course of time. Several instances of clines have already been given in this book (pp. 72, 122), as well as that of a 'reverse cline' in which powerful selection against intermediate forms produces a sharp discontinuity. It is well worth drawing attention here to a cline which has attracted considerable attention, but yet awaits genetic analysis. It is that connecting the wholly black Carrion Crow, *Corvus corone*, and the Hooded Crow, of very different appearance, having a light grey mantle. Three clines occur in different parts of the northern temperate region between areas where the birds are uniformly of one type and the other.* They seem to range between 75 and 200 km in width. Carrion and Hooded Crows interbreed and the offspring of a single pair show much variation between them, as is to be expected if the distinction is controlled on a fairly simple genetic basis.

Some account of polymorphism has already been given in Chapter 4. Here, therefore, it is necessary only to mention one or two further aspects of it that have been especially studied in wild populations. The condition is sometimes sex-controlled, but it is then not certain whether any physiological differences associated with it are operating in both sexes. The point becomes evident when we turn to the Clouded Yellow butterflies (*Colias*). This is a large genus falling

* One of these crosses central Scotland.

into two groups. In one the males are lemon yellow and the females monomorphic and whitish; in the other the males are always deep yellow but the females are of two kinds, the commoner coloured like the males and the rarer whitish. The latter are due to an autosomal gene dominant in effect, and it influences the insect's habits as well as its appearance in the females which, in their pale form, are active at a lower temperature than the deep yellow. Consequently the whitish females are relatively commoner in the early morning and late evening when samples are caught throughout the day. Though indistinguishable, the males possess this gene at the same percentage as the females. Does it influence their habits too? We do not know.

The idea of a polymorphism that does not affect visible features will have become familiar in discussing one form of incompatibility. Chemical polymorphism is indeed a widespread condition. It can be illustrated from White Clover, *Trifolium repens*, and Bird's-foot Trefoil, *Lotus corniculatus*. A number of plants give rise to hydrogen cyanide when their foliage is damaged, but in these and certain other species the reaction is polymorphic: some specimens produce the poison and others not. The distinction is controlled by a pair of genes *Ac* and *ac*, the formation of the gas being dominant. Another pair, *Li* and *li*, is involved in addition. These also are polymorphic and facilitate (dominant) or check the effect of *Ac*. Thus only the *Ac* –, *Li* – plants give rise to an appreciable amount of the cyanide.

It seems that those that do so are at an advantage because they are partially protected from defoliation by certain invertebrates that feed on them: slugs and caterpillars of various species. The matter is of practical importance, since though cattle can tolerate a considerable amount of the cyanide, a large and sudden increase in the plants liberating it can actually cause a heavy mortality in hungry animals. Sheep are less seriously affected, for it seems that they cease to feed until their symptoms disappear.

The relative frequency of the poisonous plants differs considerably from one locality to another. Generally the positives are in excess, but A. D. Jones, who has done much valuable work on this subject, has found a few places in which the non-poisonous forms are the commoner.

There is a relation between January mean temperature and the frequency of the clover producing the gas; a decline of 1°C is associated with a reduction of 4·2 per cent of the *Ac* gene and of 3·2 in *Li*. No such temperature association has been found in the *Lotus* population. It is curious that though so interrelated in their action, the two pairs of alleles concerned are not linked. The clover is the easier species to work with genetically, since it is a diploid ($2n = 32$) while the *Lotus* is a tetraploid ($4n = 24$). However, at least in northern Europe, the *Trifolium* is of the greater importance, as the *Lotus* is hardly used as a crop and fodder plant there; though it is grown for that purpose in North America.

Two exceptional situations responsible for great variability might well be omitted here but, since they occur among wild roses and brambles, their effects have aroused considerable curiosity among country-lovers. In the roses one encounters some stable species such as the Burnet Rose, *Rosa spinosissima*. This is a tetraploid, based on 7 as is the whole genus. It principally inhabits sand dunes, where it grows close to the ground. Its relatively large white blossoms with their powerful and characteristic scent, and its purple seed vessels, show it to be sharply distinct. Another tetraploid, the French rose, *R. gallica*, of southern Europe, has been important in horticulture and will be mentioned in that connection on pp. 136–7. A number of the other wild roses are pentaploids; but in western Europe there is even an apparently indigenous diploid, the Field Rose, *R. arvensis*. It is, however, in the Dog Roses, *R. canina*, that the really extraordinary rose situation is encountered: in them, indeed, the idea of species has in fact broken down.

In these Dog Roses, seven chromosome pairs give normal segregation. This is combined with different multiples of seven single chromosomes that are passed to the ovules but lost in the pollen. We have here a normal Mendelian system and a purely maternal one acting together and providing great diversity, yet limited in type. The high variability of the Brambles (*Rubus fruticosus*) is of another kind, attained simply by polyploidy.

Finally, we can turn more directly to the sights and sounds of the countryside and think of the adaptations of the Cuckoo, which must surely be mentioned here. There are a number of Cuckoo species scattered in various continents, most but not all of them parasitizing the broods of other birds, while Cuckoos are not alone in doing so. Though the European, and British, Cuckoo, *Cuculus canorus*, is not the most remarkable in this respect, it has been the most fully studied. And ever since a monk of Reading Abbey wrote the poem 'Sumer is icumin in', and no doubt long before, the call of the male Cuckoo has been taken to mean that spring at last is here. Moreover, there have always been problems posed by the brood parasitism of this bird. These, for so long canvassed, have in recent years been resolved; especially by the genius of The Hon. Miriam Rothschild and Miss Theresa Clay (1952).

The typical cuckoo-call is restricted to the male. That of the female is seldom heard because soft and less frequent; it resembles

E

the noise made by pouring water out of a narrow-necked bottle. On the rare occasions on which it is possible to listen to the male at very close quarters, it will be found that his 'cuckoo' is immediately followed by a sound similar to that made by the female, though produced more softly even than hers.

The Cuckoo lays eggs of great diversity, for the most part accurately adjusted to deceive the birds that are to act as foster parents for it. In Britain, the Meadow Pipit, *Anthus pratensis*, is often chosen for this task; its eggs are smoky brown with dark speckles. Favourite hosts also are the Robin, *Erithacus rubecula*, laying whitish-grey eggs speckled and blotched with red; and the Reed Warbler, *Acrocephalus scirpaceus*, the eggs of which are pale dull green marked with olive; and other species are sometimes preyed upon. The Hedge Sparrow, *Prunella modularis*, also much favoured in Britain, is an exception. Its eggs, which are clear blue, are not copied by the cuckoo although it is capable of laying blue spotless eggs, as it does in the nest of the Redstart, *Phoenicurus phoenicurus*, in Continental Europe. The reason, as Rothschild and Clay point out, is that the Hedge Sparrow is most uncritical and tolerates in its nest eggs that do not match its own.

It is now known that though the Cuckoo lays up to a dozen eggs, it deposits only one in each nest; also that each female lays no more than a single type of egg and selects the host suited to it. Female Cuckoos must therefore be divided into groups, known as 'gentes', according to the kind of egg they produce.

How do they choose the correct form of nest to which to carry an egg? Rothschild and Clay point out that many birds respond to sights and sounds encountered very early in life; a predilection which indeed attaches the young to the first living creature they see generally, of course, their parent. It now seems that the plumage and song of birds similar to the one which reared a particular female Cuckoo evokes the stimulus in her to parasitize the species which fostered her.

Another difficulty here presents itself. The female Cuckoo, being extremely promiscuous, must often be mated by males belonging to other gentes. In spite of this, she lays eggs of one kind only. It was Punnett who in 1933 pointed out that this result could be attained if the genes for the colour-pattern of the eggs (and in an Indian Cuckoo, their size) were carried in the Y chromosome; which is transmitted by the female in birds, though by the male in mammals. This must have been achieved by a transference, or possibly a structural interchange of material from an autosome, since the few genes for egg colour known in other birds are, in fact, autosomal. There is now some evidence that the Y chromosome is unusually large in the Cuckoo, as it might be if a piece of an autosome were attached to it.

The young Cuckoo heaves the true young of its foster parents out of the nest so as to secure the whole food supply to itself; for it seems always to be considerably larger than they. This and other aspects of the bird's habits, and the accuracy of its egg mimicry, will have been perfected by selection acting on variability supplied by the gene-complex. Indeed the selection concerned is stringent since, except for the Hedge Sparrow, the foster parents are intolerant of eggs differing obviously from their own. Any which do so they destroy or abandon.

Genetics in Cultivated Plants

This is not a book on gardening, so no attempt is made to provide a general guide to the genetics of garden plants; but rather to suggest the types of genetic adjustment which have been responsible for producing the species in cultivation. These are briefly described in something like an order of increasing complexity. With this object in view, it does not seem desirable to subdivide the material under such headings as fruits, vegetables and ornamental flowers. That would involve repetition of the same types of variation and selection. For a similar reason, lists of the genes known in each species are not provided here. Were this to be done, long tables would be needed for some of them, and these would supply but little information on the principles that have been at work in transforming wild forms into those of horticultural value.

1. Species Improved Entirely by Selection of Simple Segregants and Bud Sports

Man has relied upon normal genetic variation to improve the peach, *Prunus persica* (in which 2n = 16), and to obtain the nectarine from it. The latter is a simple and complete recessive which has arisen occasionally as a shoot on a peach tree; while the reverse, nectarine back to peach, has also been observed. A number of the features studied in the peach-nectarine group are, however, intermediate in the heterozygotes; while, of course, multifactorial variation has provided material for improvement here. The species originated in China.

We find a quite similar genetic situation in antirrhinums (2n = 16), to which polyploidy has contributed nothing to establishing this as a bedding plant. The wide range of its flower colours is due merely to the segregation of mutants.

The same is more or less true also of the sweet pea, *Lathyrus odoratus*, which in its garden form remains a diploid (2n = 14) that has gained nothing by hybridization with other species. It reached England at the end of the seventeenth century, imported from Sicily, where the flowers are uniform in appearance: small, reddish purple, and growing in pairs on short stems. It is cross-fertilizing in nature; but in England selfing has been impressed upon it by selection for rate-genes (pp. 77, 119), leading to such early development of the anthers that pollination occurs before the bud opens.

The great range of sweet pea flower colours seems due entirely to segregants; many of the genes concerned are closely linked, and some are multiple alleles. Horticulturally, a variant of much importance is the 'Spencer' form of flower, with a waved edge. It is a simple recessive, somewhat affected by modifying genes and by the gene-complex in which it is operating. Genes have been mapped on all seven chromosomes of the sweet pea.

2. Diploid Species with Artificial Tetraploids

The tomato, *Lycopersicum esculentum* (2n = 24), originated in South America. Though long eaten in Europe, where in the sixteenth and seventeenth centuries it was employed for supposed erotic purposes as the 'Love Apple', it only became known to English caterers about 1900. I can recall that before the 1914–18 war, the tomato caused surprise and comment on the lunch table.

Much genetic work has been carried out on this species, the variants chiefly attracting attention being of course those affecting the colour of the fruit, the flesh and skin of which are under separate genetic control: red flesh is dominant to yellow; and an opaque, yellowish skin is dominant to a transparent one. A combination of these two dominants is that found in the red tomato ordinarily on sale. Red flesh with a transparent skin gives a dull red fruit. Yellow flesh is modified to a deep or pale shade by these two skin types. Tetraploid tomatoes can be produced artificially. When the young plants are decapitated, a small percentage of the shoots are larger than normal and have 48 chromosomes. This, indeed, provided the first instance of an experimentally produced polyploid.

3. Simple Segregation With, in Addition, Useful Autopolyploidy

The genus *Campanula*, with its blue or white bell-like flowers, contains a number of popular border and rock plants. It supplies a suitable example here. The species basic, as it were, among those cultivated is *C. persicifolia*, a diploid in which 2n = 16. It produced an autotetraploid giant (4n = 32) 'Telham Beauty', which is a well-known horticultural success. So too is a dwarf diploid, often planted in alpine gardens. It is such a striking form that it sometimes passes under the name of *C. nitida* as if it were a distinct species; though it is a simple recessive from the normal plant.

The raspberry, *Rubus idaeus* (2n = 14), is in a somewhat similar position. It is very diverse owing, almost entirely, to heterozygosity. A number of its genes have been studied individually. The most notable variant is that giving recessive 'white' fruit. Their flavour is excellent, though distinct from the red and, unlike the gooseberry, the pale colour remains when cooked; consequently, this form can be used to make an attractive champagne-coloured jelly. There is some evidence that the white-fruited raspberry is the better crop to grow on a highly alkaline soil. In a Somerset garden of that kind, a wide selection of red-fruited strains could none of them be properly established, though no difficulty was experienced with the white. A few of the cultivated races of the raspberry are tetraploids, but they are not among the most satisfactory (see also p. 135).

4. Autopolyploidy and Divergent Selection

Here should be mentioned the sunflower, *Helianthus annuus*, a diploid (2n = 34) valuable for its seeds, and its sterile hexaploid (6n = 102) which propagates vegetatively as the perennial Jerusalem artichoke, *H. tuberosus*. This is the *topinambour* of France and has no connection with the Celestial City; 'Jerusalem' being merely a corruption of the Italian *girasole* to indicate its sunflower origin. Why the name 'artichoke' should be attached to it is hard to conceive since, of course, it is quite unrelated to the thistle genus which includes the globe artichoke.

The Jerusalem artichoke is useful to those who have to reduce the calories in their diet since its starch, being largely in the form of inulin, is not broken down by the human digestive juices, so that this vegetable merely provides bulk and a pleasant taste.

5. Evolution to a Fertile Triploid, with Polysomics

The hyacinth, *Hyacinthus orientalis*, is a diploid with 16 chromosomes as a wild plant in Lebanon, as well as in those cultivated varieties which first appeared after its introduction to Holland about 1560. All other hyacinth species have very different chromosome numbers, and have made no contribution to the forms in cultivation.

Blue is the original flower colour, and Darlington points out that most of the other shades, as well as double flowers, arose as segregants probably because of the selfing imposed upon the solitary plants in Dutch gardens. Selection then improved what rapidly became an extremely popular bulb. Forms with large flowers and leaves were naturally favoured by the growers who, of course unknowingly, were thus establishing triploids ($3n = 24$). For here it seems that the haploid condition does not comprise a sharply distinct but balanced set of chromosomes. On the contrary, it appears that each individual member contains such genes as make a self-adjusted unit. Consequently, triploid hyacinths are exceptional in being fertile; while polysomy adds or subtracts from the total so as to produce chromosome numbers spanning the range between the diploid and tetraploid values of 16 and 32.

6. Hexaploid Evolution with a Mass of Polysomics

Darlington (1963) has indicated the remarkable, indeed unique, process by which the vast array of chrysanthemums has arisen. It seems that all are based upon hexaploids from a vanished diploid at 18. The varieties have appeared both as segregants and from abnormal cell divisions, giving rise to 'bud sports' during 2,000 to 3,000 years of cultivation in China.

Today one finds the chromosome numbers ranging round their typical hexaploid value of 54 from 47 to 64. That is, what appears to be mere polysomy has at its extremes downgraded and upgraded the values almost to those that could have been reached as direct multiples of five or seven from the original progenitor in which $n = 9$. Darlington considers that the basic chrysanthemum qualities result from the presence of 45 chromosomes that have remained more or less intact, while the immense diversity of this 'species', *Chrysanthemum indicum* as it is, for convenience called, derives among other things from numerical changes in the remaining 9 (compare here the Dog Roses, p. 127). It is possible to make a generalization in regard to the result: an increase in chromosome number has gone hand in hand with an increase in flower size.

7. Autopolyploids with Useful Surviving Diploids

The ordinary strains of the potato, *Solanum tuberosum*, are mainly tetraploids ($4n = 48$), though a few are diploids. However, triploids, though sterile, are widely grown, since the tubers are relatively large. This is a Central and South American member of a widely spread group of plants in which the fruits are highly poisonous as, indeed, are those of the potato; the most famous example of this being the Deadly Nightshade.

The Plums in cultivation include the diploid Cherry Plum, *Prunus divaricata* ($2n = 16$), and the normal garden plums, *P. domestica*, which are hexaploids at 48. They are extremely heterozygous, to which their diversity is due. In the fruit, an oval shape is dominant to flattened; and yellow is dominant to red and black. The cultivated strains are subject to much incompatibility, and plum orchards must therefore be planted with due regard to a correct arrangement of the trees to secure interfertility between them. Information on this matter is available in the appropriate handbooks and from nurserymen. Some of the well-known varieties are, however, self-fertile, as is the Victoria plum; and there is, of course, no objection to planting large stands of these together.

The cherries, also highly incompatible (see p. 116), have been developed along two distinct lines. The Sweet Cherries, *Prunus avium*, are diploids (2n = 16) while the sour or morello strains, *P. cerasus*, are tetraploids. The 'black' cherries are simple dominants to the 'white', but the ornamental forms are sterile triploids. This is an advantage from the horticultural point of view, as none of their energy is expended in fruit formation.

In both apples and pears 2n = 34, while some are triploids. These latter give fewer but larger fruit which, if not carried too far, can be a commercial advantage, as seen in Bramley's Seedling and Ribston Pippin among apples, though both are highly sterile. Yet it is the great range of diploids that includes most of the popular varieties; among them Cox's Orange Pippin, which must be one of the most widely grown of all eating apples today.

There is great incompatibility among both apples and pears. Thus the planting of orchards containing them needs to be carried out with discrimination so as to bring into proximity trees of compatible types.

The tulips may be thought of as a supergenus in which n = 12. Their great diversity in flower-form and colour has arisen mainly from heterozygosity. Though many are diploids, just a few are tetraploids (*e.g.*, *T. turkestanica*); and some of the most valuable horticultural varieties are triploids: the first of these to be obtained was at an early date in cultivation, in the sixteenth century, when Keizerskroon appeared. Only in the 'Clusiana section' are pentaploids to be found, though this still includes diploids and tetraploids. The finest clump of *T. clusiana* that I have myself seen was in the Garden of Gethsemane. One of the plants bore larger flowers than the rest, and was slightly distinct in appearance. It was probably a polyploid of higher number than the others.

8. Plants from One Species Only, but Having Different Basic Chromosome Numbers

The cabbages and their allies are an exceptional group (the supergenus *Brassica*) within which the species are built up from three distinct basic chromosome numbers, 8, 9 and 10. It was only when this was discovered that, as Darlington (1963, p. 137) says, 'the past history and future possibilities of the confused groups of cab-

bages and turnips, swedes and kales became clear. The reason for
this is that different basic numbers and different stages in the polyploid series imply intersterility, and therefore determine successive
steps in evolution.' That is to say, we have here a more fundamental
distinction than that built merely upon increasing polyploid values,
as in wheat.

The Cabbage itself, *Brassica oleracea*, is a native of Europe and
southern England and, like a number of our culinary vegetables, is
a plant of the seashore and its immediate neighbourhood. It is a
diploid with 2n = 18; while the swede is a tetraploid in the same
series, in which 4n = 36, or 38 in some races owing to polysomy.
On the other hand, the turnip is built up from a different basis,
having 2n = 20. So again is black mustard, with 2n = 16. From it
is obtained the black mustard of commerce and the medically
important mustard oil.

Crosses between plants arising from those with different haploid
numbers normally give rise to sterile or semi-sterile hybrids. These
may, however, form fertile allopolyploids when their chromosome
set is doubled. For instance, the result of a cross between a swede
and a turnip (giving 2n = 28) is sterile; though on doubling its
chromosomes to 56, it forms the remarkably constant *Brassica
napocampestris*. This, being itself fertile but sterile with both its
parental forms, constitutes a new, artificially produced, species.

9. Allopolyploidy Without Subsequent Autopolyploidy

Several little strawberry species having small fruits were already
grown in gardens during the Middle Ages: one, a diploid (2n = 14),
being the Wild Strawberry of western Europe and Britain. The
others were polyploids from which the diploid had disappeared; but
they were unsatisfactory.

The great advance in strawberry culture, initiating the octoploid
garden strawberry with its large fruits, took place in France from
crossing two New World species, both octoploids (8n = 56). These
were *Fragaria chiloensis* from Chile and *F. virginiensis*, a widespread
plant in the eastern U.S.A. In neither are the diploids known. They
produced a hybrid, also an octoploid, in which the two sets of chromosomes proved to have sufficient genetic material in common to
allow pairing between them to take place. The cross so obtained
was accordingly fertile without autopolyploidy. Selection practised
upon it in England was responsible for a great advance in strawberry culture during the second half of last century. Its triumph was
the production of the famous Royal Sovereign. The diploid straw-
berries are all hermaphrodite, while the polyploids are wholly or
partly dioecious (with separate sexes, that is to say); for the occur-

rence of polyploidy tends to be associated with a change in the breeding system.

10. Species Crossing Involving an Unreduced Gamete

The origin of the loganberry has aroused much speculation. The first plant appeared in Judge J. H. Logan's garden in California about 1881. It is a hexaploid, with 42 chromosomes (7 being the basic number of its genus, *Rubus*, p. 131). On balance, it seems probable that it is the result of a cross between the raspberry *R. vitifolius*, with 56 chromosomes, of the western U.S.A. and an unreduced gamete of a diploid raspberry species, *R. strigosus*; moreover, the two were growing close together in the place where the loganberry appeared. The result would give rise to 28 + 14 chromosomes. This hybrid origin for a plant unknown in nature, or previously in cultivation, has been hotly disputed, some claiming it as merely a variety derived wholly from *R. vitifolius*. Yet experimental work seems to support the view that the loganberry arose from the species cross here indicated.

11. Allopolyploidy

Instances of allopolyploidy under experimental conditions have already been referred to in *Primula kewensis* and in *Brassica napocampestris*. Exactly the same evolutionary step, requiring subsequent selective adjustment, has repeatedly been taken in nature, and of this the Dahlias provide an excellent example.

These are Mexican plants. Diploids have not been found, and are probably extinct, all the wild species but one being tetraploids (4n = 32). They fall into two groups: the flower pigments of the one are red or yellow; of the other, magenta and of a distinct chemical type. The single exception is *D. variabilis*, the 'garden' dahlia. It possesses both double the number of chromosomes present in the others, being an octoploid, and the two distinct classes of flower pigment. This strongly suggests that it arose from chromosome doubling in a hybrid between plants belonging respectively to the two main groups. It had evolved naturally in Mexico before its cultivation, but its origin along these lines has now been confirmed by detailed experimental work.

In view of these facts, *D. variabilis* is obviously highly variable as to flower colour. It is so too in its other features, such as pattern, height and leaf shape. These qualities are basically controlled on simple Mendelian lines, but there is much complex interaction between the genes. A cross between single- and double-flowered types produces a great range from one extreme to the other on a multifactorial basis. A more intricate example of combined auto-

and allopolyploidy has already been described in the grass *Spartina* (p. 121); but this has nothing to do with garden plants.

12. Autopolyploidy with Possible Alternative Allopolyploidy

Wheat provides an example of the advantages and use of polyploidy. The primitive 'einkorn' wheat, *Trifolium monococcum*, is the original diploid (2n = 14), somewhat modified through selection by neolithic man. It is small and of poor yield, and particularly unsatisfactory because it threshes out with the glumes* attached. The next stage is provided by the relatively primitive 'hard wheats' which are tetraploids (4n = 28) that seem to be derived from a tetraploid wild species, *T. dicoccoides*; itself probably a segmental allopolyploid. These tetraploids include the 'drum wheat', *T. durum*, grown largely for making pasta, and the somewhat improved 'emmer wheat', *T. diococcum*. From the hard wheats, hexaploids (6n = 42) have probably originated, and to them belongs the bread wheat, *T. vulgare*, of commerce. However, a fertile allopolyploid from a cross between *T. dicoccoides* and the grass *Aegilops squamosa* has been produced experimentally. This closely resembled 'spelt', *T. spelta*, another of the hexaploid species; so that these may, perhaps additionally, have been derived from this type of crossing.

The Evolution of Garden Roses

It seems appropriate to end this brief survey with some reference to roses, in which so many gardeners are interested. Moreover, these plants have been diversified and improved by means of techniques already mentioned, and throw further light upon them. It will be useful therefore to treat the subject a little more fully than has been done in the preceding sections, though very shortly compared with its scope and interest.† A few words have been said about wild roses on p. 127.

The basic chromosome set in all roses derives from n = 7. A number of wild species and others grown in the East from a remote period have contributed in various ways to producing modern garden roses. They fall into two main groups: diploids from China and Japan, in the form of large bushes and climbers; and smaller tetraploid species from the Levant westwards.

Two of the eastern diploids have been of special importance and deserve particular mention. The basic one of this group is the

---

* A pair of short, stiff spikes partly enclosing the seed.

† Detailed information on the subject may be obtained from Darlington (1963), Hart (1941) and Wylie (1954).

CHINESE ROSE, *Rosa chinensis*, the ancient garden rose of China. It has departed too far in cultivation, and by selection for double flowers, varied colours and a strong scent, for its wild ancestors now to be recognized. The MUSK ROSE, *R. moschata*, has been of much importance also. It is still found wild in the Himalaya, but a form with double flowers has long been cultivated in China. The species was brought to the west in classical times, but it did not reach England until the reign of Henry VIII. *R. chinensis*, however, only arrived here in 1792.

Among the tetraploid western group, the wild FRENCH ROSE of southern Europe, *R. gallica*, has been particularly significant. It had at an early date evidently been possible to cross it with an exceptional tetraploid plant of *R. moschata* to produce the DAMASK ROSE. It was this that subsequently gave rise to the CABBAGE ROSES, *R. centifolia*, obtained by the selection of segregants during the seventeenth century, a process carried on in Holland.

It may be said that the modern roses began to appear as a result of hybridization after 1800. Though *R. chinensis* and *R. moschata* must have been crossed from remote times in the East, a particularly favourable segregant from them was obtained in the U.S.A. in 1802. By crossing it with another of the Chinese diploids, *R. gigantea*, the beautiful but none too hardy TEA ROSES, diploids as their origin suggests, came on the market.

Progress was also being made on another line. In this respect, an essential was the production of a 3n type. This occurred spontaneously in 1817 in a cross between the diploid *chinensis* and the tetraploid DAMASCUS ROSE, *R. damascina*, from the Levant: a chance event that took place in the Ile de Bourbon (now known as Réunion) and gave rise to the large-flowered and beautiful BOURBON ROSE. As expected in an ill balanced triploid such as this, it formed an occasional 2n gamete which could, and did, cross with the normal 2n gametes of *R. gallica*. Thus appeared the tetraploid HYBRID PERPETUALS.

Then, at the end of the last century, HYBRID TEA ROSES became available to the grower. They were produced from the diploid Tea and Hybrid Perpetuals, using occasional 2n gametes from the diploid progeny. As Darlington says, it was now for the first time that the chromosomes of the three great foundation stocks, *chinensis*, *moschata* and *gallica* were brought together.

Of the ten sections into which the true roses are divided, the Cinamomeae contains the greatest number of species. Yet only two of them, *rugosa* and *cinamomea*, both Eastern diploids, have contributed anything at all to the garden varieties; and even they have done so only to a very minor extent.

One very distinct advance made in the East needs to be mentioned also; the production of the POLYANTHA ROSES. They were obtained by crossing the Japanese *R. multiflora* with other eastern species, so that they are diploids. It was in these dwarf Polyanthas that a wholly new step was taken in rose evolution; the mutation to the scarlet pigment pelargonidin, previously unknown in this entire group of flowers. It gave rise to the novelty 'Paul Krampel', in the late 1920s.

The HYBRID POLYANTHAS were then obtained from a cross between dwarf Polyanthas (2n) and Hybrid Teas (4n). These were at first sterile triploids, but tetraploids have now arisen from them, largely through back-crosses to Hybrid Tea varieties.

By a very different approach, Lord Penzance originated the PENZANCE ROSES. To do so, he crossed various garden forms with the Sweet Briar, *R. rubiginosa*. This is a pentaploid species ($5n = 35$) of fairly common occurrence in western Europe including Britain, and the hybrids produced with it carry with them the Sweet Briar scent.

Darlington (1963) has pointed out that it is far easier to pass from lower to higher polyploid values than the reverse. Indeed when triploids have arisen from the cross $2n \times 4n$ among roses, they have always established themselves by adjusting to the higher rather than to the lower number. We may feel sure therefore that roses have evolved by way of their own diploids rather than from related tetraploid forms.

Finally, a consideration of the chromosome numbers in the various stocks of roses grown today indicates what types of crosses are likely to produce fertile hybrids in the future.

7. *Individual Variation in Man*

It is not always easy to draw a distinction between the diversity shown by human populations in sickness and in health. Thus the blood groups and other polymorphisms differ from one normal individual to another while, in addition, the genes responsible for them may influence liability to develop various diseases. It seems best therefore to describe a few examples illustrating these combined effects, and then to consider the more direct inheritance of illness and abnormality of various types.

The Blood Groups

Few medical techniques have been so much advertised in recent years as blood transfusion. Requests to give blood are constantly reaching the public, and the value to the community of doing so is widely stressed. Yet in so much advertisement and discussion the facts of blood grouping, basic to the whole subject, are often distorted or submerged. This is not surprising, for the matter is complex, and in my experience many of those employed in blood transfusion services have little knowledge of general serology and the genetics involved, though their work is founded upon it. There is thus an obvious need to give an account of the blood groups, of great interest in themselves, for the benefit especially of those without medical training. It will also be desirable to say something of their legal and anthropological aspects (pp. 155–7, 174–81).

At the outset it should be noticed that the numerous human blood groups fall into fifteen or so systems. Only two of these have normally to be considered in transfusion, though the existence of the others is not without effect and should always be kept in mind. Unfortunately, the two of particular medical relevance are among the more complicated, and the mistake is often made of attempting to explain them without first illustrating their plan from one of the simpler types.

Within any one system, the groups are controlled genetically as a polymorphism: that is to say, by genes or supergenes determining clear-cut alternative types, and it will be appreciated that they must

be maintained in the population by balanced advantages and disadvantages (pp. 78–9). But before that situation can be explained
in any detail, a few words must be said about the structure and
functions of human blood.

This consists of a nearly colourless liquid, the plasma, which
transports some of the products of digestion, hormones and other
substances; also much of the carbon dioxide to be excreted from the
body, but in the form of sodium bicarbonate. In the plasma float
separate cells of two main kinds: the red corpuscles, or *erythrocytes*,
and the white corpuscles or *leucocytes*. The red corpuscles contain
haemoglobin, the compound of iron which gives blood its colour. This
is a dark red, but turns scarlet when transporting oxygen, which is
its function. These cells are formed in bone marrow and need constant replacement. They cannot survive for long, five to ten weeks
only, because they are incomplete: that is to say, they lose their
nuclei before being passed out into the blood. This is not normal
in the vertebrates, in which the red corpuscles are always nucleated
except in the mammals. Indeed, their non-nucleated condition can
nearly be taken as characteristic of the class; but not quite because
they are nucleated in the Monotremata,* the most primitive
mammals of all.

The white corpuscles contain no haemoglobin. They are of several
kinds and, though some are larger than the red, they are much less
numerous; in the proportion of about 1 : 600. All are nucleated, and
one type destroys bacteria and foreign bodies. They are diverse in
origin. Some (lymphocytes) are formed in the lymph glands scattered about the body; these glands include the tonsils. Others
(e.g. polymorphs) arise in the bone marrow.

In addition to the corpuscles, minute particles known as *platelets*
float in the blood. They are fragments of cytoplasm derived from
large cells in the bone marrow. In combination with *fibrinogen*, a
complex protein in solution in the plasma, they give rise to a microscopic network of fibres which entangles the blood cells and forms
the blood clots that prevent excessive bleeding. If a sample of blood
is allowed to stand in a tube, a clot separates out in it and sinks to
the bottom, leaving a straw-coloured fluid above. This is known as
serum, and is simply plasma after the fibrinogen has been removed
from it.

We can now turn to the human blood groups. The red corpuscles
carry on their surface proteins known as *antigens*, each of which
interacts with a corresponding *antibody*, if present, in the plasma.

---

* The duck-billed platypus, *Ornithorhynchus*, and the two genera of spiny
anteaters, *Echidna* and *Pro-echidna*, all restricted to the Australian region.

Serious effects follow when this happens. It is obvious, therefore, that corresponding antigens and antibodies cannot normally co-exist in the same person. They can, however, be brought together as a result of a transfusion; also of a pregnancy.

The presence of a given antibody may be spontaneous, resulting from the genetic constitution of the individual, or *immune* when it is produced as a reaction stimulated, for example, if inappropriate blood is used in a transfusion. The introduced red corpuscles then 'agglutinate' or clump together in little masses which may block the finer vessels. This, though formerly thought to be their chief danger, is probably not very important, because the agglutinated cells undergo haemolysis: that is to say, they are broken down, releasing active substances into the circulation. The pigment thus liberated is rapidly excreted, so that the urine turns black. That condition is accompanied by constricting pains in the back and chest and by shock symptoms: sweating, shivering, nausea and an extreme increase in pulse rate which can be severe enough to cause death; especially in those debilitated already by loss of blood; as they will be when a transfusion is required.

Though, as already mentioned, blood-group incompatibility usually arises in respect of two systems of special practical import-ance (to be described on pp. 144–8), it can also be due to one of the others. Furthermore, it can also result from a second transfusion with the blood that was satisfactory when used the first time (p. 147). Consequently, a test for compatibility should, if possible, always be made before a transfusion, even if someone who on general grounds should be an appropriate donor (pp. 144–5, 147) is available.

This is easily performed. A minute drop of the recipient's serum is placed on a white glazed tile and a similar drop containing the corpuscles to be tested, diluted (with the correct salt solution) to reduce their relative number, is placed alongside and then mixed with it. If agglutination takes place, the clumping can usually be seen with the naked eye, or at any rate with a (× 10) hand lens. This represents the main plan of such a test, though various minor but important details have to be observed in carrying it out.

For a number of years after the blood groups were discovered in 1900, clotting during transfusion proved a great difficulty. This was overcome when it was found that sodium citrate acts as an effective anticoagulant, and is harmless at the concentration re-quired. Many other substances are now known to be satisfactory for this purpose also.

Notation

Serologists have not always been geneticists, nor realized that there is a standard notation in genetics from which the blood-group genes cannot be exempt. This has led to mistakes even in recent years. In 1968, for example, Race and Sanger denoted the gene for the then new Sd blood group as Sd^a, which tells us that it is recessive in effect; and then proceeded to state that its effect is dominant.

But far worse, serologists have not even co-ordinated their usage among themselves, representing quite similar conditions in entirely different ways, to bemuse those who attempt to understand them. Nor has the need been realized for a standard plan by which genes, antigens and antibodies can be interrelated consistently. The notation introduced by myself (Ford, 1955), though doubtless capable of improvement, overcomes these defects; moreover, it is so devised as to accommodate future discoveries. It must be outlined at this point so that various aspects of blood grouping can easily be discussed here and in the next chapter of this book.

Genes: symbols for genes are to be in italics. Those for antigens and antibodies are never italicized.

We are to think about any blood-group gene in two stages. First, its locus on the chromosome, which is to be shown by an upper-case letter or a pair of letters of which the first is in the upper-case (K or Lu). We have then to consider the alleles at that locus. These are to be represented as the first two letters of the alphabet, more if there be multiple alleles, placed as a suffix (K^AK^B); using a capital when the effect is dominant and a small letter when recessive. Here we meet the unusual condition that both alleles at a locus can be dominant and therefore exercise their effect even in single dose, though this may be somewhat greater for the homozygote than the heterozygote. That is to say, K^AK^A will produce one type of antigen, and K^BK^B another; while K^AK^B will produce both. If we were to find a further allele, we should represent it by the third letter of the alphabet, using the capital if the antigen it produces behaves as a dominant, and a small letter if it behaves as a recessive. In the latter event, K^cK^c would give the third type of antigen; while K^AK^c would give only that due to K^A.

An antigen cannot be detected except by interaction with a corresponding antibody, and this may be rare even when its antigen is common. Sometimes no such antibody has been discovered, perhaps it does not exist; and therefore its antigen, if there be one, is not known. We then omit the suffix. Thus, for instance, we have the genes Xg^AXg, here represented in the heterozygous state. The homozygote $XgXg$ has no known effect: but there is a locus symbol ready here for use if an effect due to the gene in question be discovered as, in fact, has sometimes happened in similar circumstances. Of course it will be realized that from the genetic point of view, Xg^AXg^A and Xg^AXg are distinguishable, owing to the types of offspring to which they can give rise. The effect of a gene without a suffix is recessive to one with a capital letter in the suffix, but dominant to that produced by one represented with a small letter in the suffix.

Antigens: the locus symbol and suffix letter, unitalicized, are used to distinguish the antigen.

The antigen group is shown by the locus symbol; and the suffix letter, following within brackets, indicates the type of antigen present. When an antigen is dominant the capital, of course, appears for the letter within the bracket; while the small letter is used for the recessive. Plus or minus signs follow the letters in the bracket to show the presence or absence of the antigen. Thus, the genes for the Duffy Group are Fy^A and Fy^B. The antigens, therefore, are Fy(A +B +) or, for instance, Fy(A —B +). If we wish simply to refer to an antigen, the plus or minus sign is unnecessary: one need merely speak of the Fy(A) antigen.

Antibodies: these are indicated by the locus symbol, with the prefix 'anti' hyphened to it, followed by the allele letter in brackets: for instance anti-Fy(A).

Examples of the Blood Groups

It has already been mentioned that only the two best known and most important of the blood-group series, with one of the simpler type to introduce them, are to be described here. Their relation to disease will be mentioned on pp. 148-9. Of the fifteen or more blood-group systems known, one is sex-linked while the rest are controlled autosomally by major genes or supergenes. A number of other apparent blood groups have not yet been fully studied. Some seem to belong to distinct systems not so far clearly understood, while others are of doubtful status.

Here, then, is a simple blood group to illustrate its main features:

The Kidd System
genes: Jk^A, Jk^B, Jk
antigens: Jk(A +B +), Jk(A +B —), Jk(A —B +), Jk(A —B —)
antibodies: anti-Jk(A), anti-Jk(B)

The frequencies of the three best known genotypes in England are Jk^AJk^A, 24·8 per cent; Jk^AJk^B, 50 per cent; Jk^BJk^B, 25·2 per cent. $JkJk$ is a rarity. Thus about 74·8 per cent of the population belong to the antigen group Jk(A) and 72·2 per cent to Jk(B). Anti-Jk(A) sera have been found many times. They have generally been immune, but a few have been spontaneous. The anti-Jk(B) serum is very much rarer. Both have given rise occasionally to serious reactions on blood transfusion.

The ABO System

The blood groups were discovered by Landsteiner in 1900, and those of the ABO system were the first to be found. For this there is a good reason, for their antibodies, instead of occurring quite rarely as with most groups, are invariably present if compatible. Consequently, transfusion from an unsuitable donor necessarily leads to dangerous agglutination. On that account, a mixture of blood from two individuals gives one or another of two clear-cut reactions: either the red corpuscles remain separate or they coalesce in small clumps. It is to detect this distinction that blood grouping is undertaken, and therefore it is primarily concerned with the ABO system.

This comprises four principal types, owing to the existence of two main antigens. These can both be absent (Group O), one or the other can be present alone (Groups A and B) or both can occur together (Group AB). That situation should be clear from the following notation:

groups: O   A   B   AB
genes: GG $G^A G^A$ or $G^A G$ $G^B G^B$ or $G^B G$ $G^A G^B$
antigens: G(A—B—)   G(A+B—)   G(A—B+)   G(A+B+)
antibodies: anti-G(A+B) anti-G(B)   anti-G(A)   absent

The percentage frequencies of these blood groups in southern England are: O, 43·5; A, 44·7; B, 8·6; AB, 3·2. They are, however, widely different in some other populations (pp. 174–81).

The possibilities of transfusion are set out in Fig. 7.1 below: these, in one form or another, though not quite in this one, have often been published.

recipient

		O	A	B	AB
Donor	O	++	+	+	+
	A	–	++	–	+
	B	–	–	++	+
	AB	–	–	–	++

++ = safe, + = probably safe, – = not allowed

Fig. 7.1.   Transfusions relative to the ABO system.

Those who belong to Group O are known as 'universal donors' since they can, with reasonable safety, give blood to recipients of any group, though they can receive blood from no group but their own. The reverse is true for AB.

This information can easily be derived from the notation just given if an additional consideration is borne in mind. An individual may not act as a donor if his red corpuscles can be agglutinated by the plasma of the recipient. It is, however, generally held that the reverse procedure is harmless: that of giving blood which will agglutinate the red corpuscles of the recipient; for the plasma so received is generally too dilute to produce serious results. That is why, for instance, universal donors can generally give blood safely to groups other than their own. Yet it is doubtful if that technique is quite free from risk. On the whole, it is best to transfuse between individuals of the same blood group. Considering the rarity in most parts of the world of group B, and always of AB, this may not be possible in respect of them. Provided time and opportunity allow, the donor's blood should always be tested against that of the recipient, and on each transfusion (p. 147). This will in addition overcome any danger due to other blood group systems, which certainly exists even though their antibodies are, at least individually, rare.

It has been mentioned that several minor subdivisions of the ABO group have been discovered. Only one of these need be mentioned here, owing to its interest in anthropology (p. 174). The reaction of the G(A) antigen may be strong (Group A1) or weak (A$_2$). The distinction does not appear to affect transfusion. The multiple allelic series at the controlling locus therefore takes, in full, the form G^{A1}, G^{A}_2, G^B, G. We are here involved in a corresponding subdivision of the antibody. It will be appreciated that group AB can, in reality, be subdivided as A1B and A$_2$B. The G(A1) antigen is dominant to G(A$_2$).

Linkage has been discovered between the G locus and a rare gene, recessive in effect, responsible for abnormal fingernails and absence of the kneecap. The cross-over value is 12.

And now we encounter a mystery. Not surprisingly, for obviously one cannot assume that every unexpected feature has already been resolved. It is astonishing that several of the human blood group antibodies are found in large quantities in certain plants. For instance, extracts of the seeds of the broom, *Cytisus sessifolius*, agglutinate O and A$_2$ erythrocytes, but those of no other blood groups, not even A1; but for this the labiate plant *Hyptis suaveoleus* contains the antibody. Various instances of the kind could be quoted. Nor do ABO antigens alone occur in this way; those of some other blood groups do so as well. Why are they found in plants at all and, seemingly, in a few species only, chiefly but not entirely Papilionaceae (the family containing peas and beans)? We simply do not know.

The ABO antigens on the erythrocytes are present also in very small amounts in certain, but not all, of the body fluids; urine and tears, for example, and in very large amounts in saliva and semen. Into these latter they do not merely leak but are actively secreted, being present in higher concentrations than in blood. This situation depends on the action of a gene *S*, dominant in effect, which allows the conversion of the antigens, always alcohol-soluble, into a water-soluble form. Its allele, when homozygous, *ss*, prevents this, and then these substances are restricted to the erythrocytes. About 22 per cent of Europeans are of this latter type, but the proportion is very different in some other races (pp. 175, 179). However, blood grouping by way of the saliva instead of blood is not satisfactory, as it is difficult to distinguish between group O and non-secretors.

There is linkage, with a cross-over value of 9, between the Secretor gene *S* and that for another of the blood groups, the Lutheran. The interesting possibility arising from that situation will be mentioned on p. 151.

The Rhesus System

This is the other blood group that is of outstanding importance in medical practice. Unfortunately, it is the most complex of all. Its curious name relates to the way in which it was discovered; for its antibodies were originally produced by injecting the blood of Rhesus monkeys into rabbits. The groups included in it are controlled by three loci which have been built into a supergene. Their effects show almost complete dominance. In antibody formation they fall in a sense between ABO in which these substances are always present if compatible,* and the rest of the blood groups in which they are rarely formed. For in Rhesus they may be frequently and easily produced as immune reactions, though they are rarely spontaneous.

The basic features of Rhesus are indicated by the following notation. Two alleles are well known at each of the three loci of the super-gene, but rare additional ones have also been detected at each site. We have then three sets of alleles, as follows:

genes: C^A, C^B; E^A, E^B; D^A, D.
antigens: C(A) and C(B), E(A) and E(B), D(A).
antibodies: anti-C(A), anti-C(B); anti-E(A), anti-E(B); anti-D(A).
  The *D* locus does not lie between *C* and *E*. The various percentages differ considerably: C^A, 42; E^A, 15·5; D^A, 59.

* The P system, not discussed here, also includes such a type.

Men and women are either Rhesus-positive or Rhesus-negative, depending on whether or not they carry the antigens which, when injected into a person lacking them, may give rise to the corresponding antibody formation. Actually, only D^A in either genotype (D^AD^A or D^AD) and homozygous C^A and E^A do so with sufficient frequency to be classed as Rhesus-positive. However, anti-D(A), which can be stimulated to arise in the DD genotype, is so much the most frequent of these antibodies that it is responsible for about 95 per cent of the instances in which they occur. In fact, we should not be far from the truth if we say that the Rhesus-positive condition is almost confined to the D(A) type. In Britain about 15 per cent of the population carry the genes DD, C^BC^B, E^BE^B, and these are the decisive Rhesus-negatives. About 85 per cent of the population are Rhesus-positive in most European countries.

It is essential that an individual with the genes DD should not be injected with blood containing the D(A) antigen, as this will stimulate the formation of anti-D(A), which persists indefinitely in the circulation. A second transfusion of Rhesus-positive blood into a Rhesus-negative individual, though satisfactory when first used, can then produce a serious reaction. We can easily picture the, sometimes disastrous, advice often given before the Rhesus series was understood: 'Let the patient have another transfusion of the blood which did him good before.'

There is a further source of danger in this respect. That is to say, a Rhesus-negative (genetically DD) woman may be pregnant with a Rhesus-positive foetus, which has received the gene D^A from its father. There is very often some escape of foetal red blood corpuscles into the mother's circulation at or near delivery, and such Rhesus-positive foetal blood can after some months stimulate the antibody in her. This may make its way back into a subsequent foetus, giving rise to dangerous haemolytic disease in it. For the interaction between the corresponding antigen and antibody may cause much destruction of the child's erythrocytes. This produces jaundice, often serious enough to prove fatal; it may also lead to brain damage.

Such disease in a newborn child, liable to appear at a second birth, will then affect all subsequent children unless special steps are taken to avert it. Parents who have once had such an affected infant cannot normally expect to produce a healthy baby again.

Actually, these sinister results are a good deal less frequent than might be expected. In western Europe, the father is Rhesus-positive and the mother Rhesus-negative in about one marriage in eight, while only one in 150 to 200 infants are affected by the disease; though this constitutes a vast problem in any country as a whole. Evidently, therefore, some protective mechanisms are operating here. Several of these are known and one of them may be mentioned in this account: owing to an appropriate evolution of the gene-complex, about 30 per cent of people are incapable of making Rhesus antibodies no matter how great the stimulus.

While several methods of treating haemolytic jaundice due to Rhesus have been established, Sir Cyril Clarke has devised a brilliant and largely effective means of *preventing* the Rhesus immunization of the mother. For she can now be given an injection* which, if received within seventy-two hours of delivery, masks the antigen on her foetal red cells and so safeguards her next infant.

One linkage relation with Rhesus has already been detected (the C.O.V. = 10 to 15). This is with the gene, dominant in effect, giving rise to elliptical red corpuscles. Its harmful effects are not known. It is rare in man and all other mammals except the camel, in which it is normal. A second rare gene, similarly distorting the shape of the human erythrocytes, is not linked with Rhesus; but is responsible for slight anaemia.

As already suggested, the frequency of the Rhesus genes varies greatly from one race to another; in consequence, so does the proportion of Rhesus-negatives in the population. This matter will be further mentioned in discussing racial differences in mankind (pp. 174–81).

Human Polymorphism

When in 1940 polymorphism was defined, it became clear that much human variation is of this kind. Moreover, in 1942 I drew the inevitable conclusion that the blood groups are polymorphic, and that consequences of importance must follow from that fact. The genes and supergenes controlling them have no obvious effects upon the body, nor upon choice in marriage; thus they must affect fertility, general stamina or health. Consideration along these lines led to the prediction in 1945 that the blood groups and other polymorphic qualities must be associated with susceptibility to develop specific diseases.

Six years after that conclusion was published, the first instance of the kind was reported. Such associations are now progressively coming to light. Here it will be reasonable to quote just a few of them.

The tendency to develop cancer of the stomach is higher among those who belong to group A than to the other groups of the ABO series. Sir Cyril Clarke and his colleagues have shown that duodenal ulcers are increased relatively by about 35 per cent in group O, and

* Of 100 μg of 7S anti-Rh gammaglobulin.

that this effect is still more considerable, by about 40 per cent, in those who do not secrete the A and B antigens into their saliva.

Even when originally predicted, infections were included in the association between polymorphism and disease. A throat infection by a haemolytic streptococcus (type A) can lead to rheumatic fever (which, however, has lately become a rarity at least in Britain). Those who belong to group O, and those in whom the A and B antigens are water-soluble, are significantly protected against that event. We can here refer to a matter of much wider consequence: that is to say, those who belong to groups A and AB are much more likely to develop smallpox, and to do so in a more severe form, than are those of the O and B blood groups (Vogel and Chakravartti, 1966). The association seems restricted to unvaccinated populations during severe epidemics. It must in the past have been of immense importance in human affairs.

Most, but not quite all, of these associations between the blood groups and disease can have had but little significance in balancing the polymorphism involved; but even so they are a pointer to the fact that the genes concerned have other and important effects in addition to the feature by which they are normally recognized. Of this, indeed, we have powerful evidence. Thus there is differential elimination of incompatible blood-grouping types during pregnancy. This actually amounts to 10 per cent against a group A foetus in a group O mother: the two commonest of the ABO types.

Since polymorphism must be maintained at balanced frequencies in the population by opposed advantages and disadvantages, the latter will often include liability to disease, and this will not be restricted to the polymorphism of the blood groups. For instance, it would be difficult to imagine a more trivial quality than the mere ability to taste a particular sulphur compound, phenyl-thio-urea; in fact no one ever had tasted it before this century, when it was first synthesized. Here we have a gene, dominant in effect, which allows about 70 to 80 per cent of the population in western Europe to detect this substance. It is intensely bitter to those who can do so but tasteless, like chalk, to those who cannot. Evidently, from its frequencies, this is a polymorphism, but we do not yet know what its contrasting advantages and disadvantages may be. That they exist, and that the genes involved have important effects, is nevertheless shown by the fact that the two alleles are respectively associated with alternative forms of thyroid disease.* This cannot be

* The thyroid is a gland in the front of the neck which deals with the iodine requirements of the body. Its alternative defects to which reference is here made are Graves' disease, much commoner in tasters of phenyl-thio-urea, and adenomatous goitre, commoner in non-tasters of that substance.

the main selective feature, but it is a straw that shows the way the wind blows.

Many of the ordinary and obvious variations in man are polymorphic. This is partly true of eye colour. Blue is recessive to the brown shades, the diversity within which is mainly multifactorial.

The gene for blue eyes must be maintained in those populations in which it occurs by balanced advantages and disadvantages, but we have no indication what these are. That qualification is, however, not quite true of another striking human attribute, red hair which is combined with a fair skin. This is recessive to black and brown hair though not infrequently the heterozygotes are slightly intermediate. The fair-skinned red-haired type is, of course, particularly susceptible to the effects of sunlight. During the 1939–45 war, the army had apparently made no effort to recognize this fact, and red-haired men were constantly sent out to India and elsewhere, whence they had soon to be returned owing to violent sunburn which would wholly incapacitate them. This liability is an obvious outcome of the pigmentary condition itself, while the gene responsible for it has a second effect: the production of freckles. This, unlike the red hair, is dominant; also it is affected by the environment, being accentuated by exposure to sunshine.

However, we have some indication that the gene has another effect of a very different kind. It has been found in a mental hospital in Australia that the proportion of red-haired inmates was to a highly significant extent lower than in the normal population from which they came (Nicholls, 1969). This suggests some advantage while, on the other hand, red-haired men and women have an increased susceptibility to skin cancers, especially in the strong Australian sunlight.

There are in fact several types of the red-haired condition, but the genetic distinctions between them are not known; they could well be multiple alleles. Some, but not all, have an unpleasant odour to many normal people, a fact perhaps concealed from them at least by their best friends. Another, or very probably the same, variant can be detected in a marsh where the day-flying Burnet moth, *Zygaena trifolii* (p. 85), is common. These insects mob certain red-haired people but take no notice of others, nor of those whose hair is not red.

On turning back to the general account of the subject on pp. 78–84, it will be obvious that in polymorphism we will often encounter the special features to which that form of variation is likely to lead: the establishment of supergenes and of heterozygous advantage. These have already been detected in a number of instances in man.

As already explained, the Rhesus groups are controlled by a

supergene, and in this connection one may especially mention the MNL, Hunter, Henshaw System; though it is not discussed here. But it has been pointed out that the end result of supergene formation is to hold distinct loci so closely together that, as in the ABO series, they resemble multiple alleles; for which indeed they have at first repeatedly been mistaken. Thus it may be said that the existence of a multiple allelic series in polymorphism may generally be taken to indicate a supergene.

It is noteworthy that the genes for the secretion of the ABO antigens into the saliva, and for the Lutheran blood group (one that is not described here), both associated with blood grouping, that is to say, are fairly close to one another on the same chromosome: in Britain, there is 9 per cent of crossing-over between them. This may merely be a chance occurrence but, as both of them are polymorphisms, a study of their linkage values in other and very different races (say, Japanese and Bantu) might show something significantly different from it; so actually indicating that the evolution of a supergene is here taking place in man.

Heterozygous advantage is the usual method for maintaining polymorphic variation; and its effects, direct as well as indirect, are evident in human genetics. They should be clear in the A and B groups of the ABO series, but a search for them cannot yet be carried out because no means of distinguishing those individuals that are homozygous or heterozygous at these loci (G^AG^A from G^AG, for example) has yet been found. However, when two members of the AB group marry, all three genotypes (G^AG^A, G^AG^B, G^BG^B)are produced and are distinguishable among their offspring. It is then evident that the heterozygotes are present in excess.

Among the numerous chemical variants now recognized as polymorphic in the human body, the haptoglobulins may be selected for brief comment, owing to the light they throw on genetic action. These substances affect the disposal of haemoglobin when released from worn-out red blood corpuscles, and are to be detected only by chemical means (electrophoresis). They are controlled by two pairs of alleles without dominance. One, Hp^1Hp^1, produces a single type of haptoglobulin, while the other gives rise to a number. The heterozygotes, Hp^1Hp^2, produce a further series differing from those due to the homozygotes: an illustration of the fact that the genes may interact to evoke their results.

The haptoglobulins have no visible effect upon the body, but the subject of polymorphism can usefully be carried further in an example to be studied by a microscopic examination of the blood; for this opens up an important field of enquiry.

Various types of haemoglobin are known to exist in man. The

first to be formed during development is *foetel haemoglobin*, HbF. It still predominates at birth, but has been wholly replaced by the normal *adult* form, HbA, four months later. The latter consists of two parts: a major one A1 and a minor A2. A mutant, restricted very largely to negroes, changes the A1 component to an abnormal type: the sickle-cell haemoglobin, HbS. This is so called because it gives rise, when homozygous, to a condition in which the red blood corpuscles become sickle-shaped, with thin processes extending from them. It leads to anaemia that is fatal at any rate by the age of 25, but often in childhood. In such people even the adults possess slight but variable amounts of foetal haemoglobin that ought to have disappeared in the first year of life. These homozygotes, then, have the HbA2 and HbS haemoglobins, with a variable trace of HbF.

The heterozygotes, on the other hand, have also the full quota of HbA2 as well as half the correct amount of HbA1, the rest being replaced by HbS. Their red corpuscles are of the ordinary shape when in circulation, their supply of HbA1 being enough to ensure this. Yet it is not sufficient to ensure it if oxygen be excluded from a drop of blood; for then the red corpuscles become sickle-shaped, but less markedly than in the homozygotes. Exclusion of oxygen therefore provides a test to distinguish normal blood from the heterozygous sickle-cell type.

The HbS haemoglobin, even in the half amount possessed by the heterozygotes, confers powerful resistance to malaria in its severe tertian form (due to the parasite *Plasmodium falciparum*). This gives them a great advantage in regions where that disease is prevalent; though ordinarily they must be somewhat handicapped, since they are not found where it is absent. Surprisingly, as much as 40 per cent of some African tribes, to which the condition is principally restricted, are heterozygous for the sickle-cell trait, even though when homozygous it leads to an early death; moreover the homozygotes rarely reproduce.

A useful comment on dominance and recessiveness may be made here. In affected individuals the disease is recessive, while the formation of HbS is not. Moreover, the protection extended by the haemoglobin is dominant. It will be recalled that dominance and recessiveness are properties not of genes but of the qualities for which the genes are responsible (pp. 21, 100).

A remarkable situation, that of benign sickle-cell anaemia, has now been discovered. It has been found in several oases in Arabia, where many of the inhabitants are undoubted homozygotes since it can be shown that their HbA1 is absent. They have HbS, proved identical with the African, the normal amount of HbA2 and a considerable but very varied quantity of foetal haemoglobin, with an average of 19 per cent. Yet the severe sickle-cell symptoms of the homozygotes have in them been almost entirely overcome: their health is nearly normal, they live to a considerable age and reproduce successfully. It is clear that they are able to do so by utilizing their persistent foetal haemoglobin. As already mentioned, some trace of this exists in the African homozygotes, and selection has evidently operated on the variable quantity found in them to produce homozygotes that are both viable and protected against the malaria.

A further form of such protection brings us back to the association between blood groups and disease (pp. 148–50), for it has now been shown that a malarial condition due to a different parasite, *Plasmodium knowlesi*, is common in west Africans if they are positive for the A or B antigens of the Duffy blood-group series. Those who lack them are protected, and consequently the Duffy-negatives are at a high frequency in the affected region; and there the monkey population acts as a reservoir for this *Plasmodium*. The Duffy system is controlled on somewhat similar lines to the Kidd (pp. 143–4) but, having certain complications, it was not chosen as a basic example of the blood groups.

It has sometimes been said that when the three genotypes are present in their expected proportions of $p^2 : 2pq : q^2$ (p. 107) the situation is not polymorphic. This can be a mistake. For when the least common of them occupies, say, 2 per cent of the population or more, it might well be asked how it can have reached such a frequency if the numbers be large. It cannot have done so by mutation which, inevitably, is far too infrequent (p. 156); nor by the spread of a gene neutral in survival value compared with its allele, for that process would be immensely too slow (p. 56). But if it be due to selective advantage of the rarer gene, what has prevented this from spreading further at the expense of its allele? Additionally, in the majority of the blood groups, as in many other comparable human situations, it can hardly be thought that such a change is now in progress (transient polymorphism). The obvious answer is heterozygous advantage but, if so, why are not the heterozygotes in relative excess? This difficulty has so affected the thought of some serologists that they have actually set down certain blood groups, in which the heterozygotes are apparently not above their expected proportion, as 'non-polymorphic' (Race and Sanger, 1976). How they think the rarer gene has reached its present distribution is not apparent. But we all know the attitude 'Set aside the more difficult aspects of a problem, and then the simpler ones can be explained.'

In man, failure to detect heterozygous advantage has hardly been considered, as it ought to be, in relation to age; especially that of foetus to adult. However, in this matter a valuable comparison is provided by the few instances in which that state has in other species been studied throughout life.

Fujino and Kang (1968) found that in the tuna fish, *Katsuwonus*, the genes controlling a physiological polymorphism, that of the substances known as 'transferrins', is subject to marked heterozygous advantage in the young which disappears in the older specimens. When plotted against time, the three lines representing the observed divided by the expected frequencies of the genotypes are well separated in the young, with excess heterozygotes, but they converge and meet in the older animals. As the authors reasonably say, heterozygous advantage is here gradually eliminated by opposing differential viability during life. Considering, moreover, that though so little studied, powerful differential elimination of blood groups before birth has already been discovered in man (p. 149), the disappearance of heterozygous advantage during life in the tuna should throw light upon the absence of that condition in certain adult, and indeed post-embryonic, human polymorphisms.

It is curious how much resistance is still found to the idea of natural selection. It chanced that some of the first examples of an association between human polymorphism and disease related to disorders of the alimentary canal.* Consequently, those who oppose the general importance of selection suggested that such interrelation is restricted to disease of that one part of the body. Even when the incorrect working of the thyroid was involved, an attempt to salvage that particular argument was made by pointing out that this is a gland which in early embryonic development is derived from the upper part of the tube leading from the mouth to the stomach (in fact, the 'pharynx'). However, such a limitation could not be sustained in respect of smallpox. In a further attempt to minimize the impact of the blood groups upon disease, it was maintained that only the ABO series was known to be involved. Now that such an association has been demonstrated in respect of the Duffy blood groups, that particular criticism has foundered also.

In general, it must be realized that when two or more distinct and genetically controlled features occur together, the rarer occupying at least 2 per cent of the population, such variation must be polymorphic, with all that this involves. Thus, however trivial may be the features by which we distinguish it (the ability to taste phenyl-thio-urea or to smell freezia blossoms), the condition must be

* The digestive tract and the tube leading to it.

an important one; being subject to selective advantages and disadvantages, generally powerful. Consequently, in discussing the significance of polymorphism, the distinguished physician Sir Cyril Clarke pointed out (1964, pp. xi–xii) that once its scope had been mastered, 'ready access to big hospital populations means that any common character can be investigated on a large scale.'

The Legal Aspect of Polymorphism

Polymorphism can often contribute decisive information on identity and parentage. This is due to the relatively high frequency of even the rarer of its phases; also to the nature of their control, which is simple in action even when complex in structure. Yet certain requirements must be met if polymorphic variation is to be used in this way. Its genetics must be thoroughly understood and the distinctions involved must already be apparent in early infancy and maintained into old age. They must, moreover, be independent of environment, including disease.

All polymorphisms which satisfy these conditions are potentially valid for legal purposes of this kind. For instance, it would be highly suspicious if a pair of blue- or grey-eyed parents had a child with dark-brown eyes, though the reverse would be quite normal (p. 150). However, the blood groups provide by far the best evidence of parentage. For those most likely to be used are based upon a vast mass of information, and the distinctions involved are objective. However, the required antisera must be easily and widely available, as are those of the ABO series; and furthermore, it is essential that the necessary tests should be carried out by an expert.

The blood groups, then, may give a clear decision upon two legal issues: first identity, when, for instance, it is thought that babies may have been accidentally interchanged in a maternity hospital; and second in cases involving disputed paternity. Indeed it has been necessary to turn to them for evidence in an immense number of such disputes in the U.S.A. and in European countries.

Two aspects of blood grouping can be called upon in the latter situation. First, paternity can be excluded if a man and the mother both lack an antigen present in the child in question. Second, a man cannot be the father of a particular child if antigens which he must hand on are absent from it. One or two exceptions to these statements involve a few of the blood groups; but these are quite well known, and such groups need not be used. There is, however, one universal source of error which cannot directly be excluded: that is to say, mutation. However, as already mentioned, a mutation-rate of 1 in 80,000 individuals seems to approach the upper limit, while no blood-group gene is in the highly exceptional position of lying close to the heterochromatin (p. 56). Indeed the average frequency seems to be about 1 in 1,000,000; and Race and Sanger (1969, 5th edn.) with their great knowledge of serology, attribute something like this value to the mutation of all the blood-group genes. Consequently, mutation can be excluded if two distinct blood-group loci be employed; for the chances that two coincident exceptions are mutational is then in the neighbourhood of one in 1,000,000 × 1,000,000. True, a very distinguished lawyer said to me recently: 'Yes, but an exceptional type of child might represent just that occurrence.' So it might, but it is certainly true to say that no legal decision has ever been reached in which the possibility of error is not greater than this. If we are really to take into account chances of such astronomical rarity as one in 10^{12}, all courts of law must now close.*

When applying the results of blood grouping to obtain information on identity and paternity, the ABO system must be the first choice possessing, as it does, every one of the desirable attributes already mentioned. Its use is shown in Fig. 7.2:

type of marriage	blood groups absent among children
O × O	A, B, AB
O × A	B, AB
O × B	A, AB
A × A	B, AB
B × B	A, AB
A × B	———
O × AB	O, AB
A × AB	O
B × AB	O
AB × AB	O

Fig. 7.2. Blood groups of the ABO series that cannot appear among the children from known types of marriage. (Reproduced from Ford, 1973, with the kind permission of Chapman and Hall Ltd.)

It will be noticed that there are many situations in which it provides no relevant evidence. Additional blood groups must then be called into service, as they must also in order to exclude any possibility of error due to mutation.

* 10^{12} is an immense number. In terms of time, it is actually more than half the number of *seconds* from now back to the battle of Marathon (490 B.C.).

It is held that the A1, A2 subdivision is not sufficiently accurate for legal work. However, Rhesus can often provide valid information in these circumstances; and an inspection of its genotypes (p. 146) will indicate which blood groups must be absent from a legitimate child. Moreover, evidence from the numerous blood group series not mentioned in this book is widely admissible: for information on that aspect of the subject, see the authoritative account given by Race and Sanger (*l.c.*). It is interesting to note that the presence or absence of ABO antigens in the saliva can now be determined with sufficient accuracy to provide legal evidence, and that this is possible by tests using extracts of the gorse plant, *Ulex* (p. 145).

Inherited Diseases in General

Numerous diseases and defects, known to be inherited, are not polymorphic (Ford, 1973). Only a few important instances of these can be given in this book. A list of them, approximately complete up to the mid-1960s, can be obtained from McKusick (1966). Some are multifactorial. These are necessarily the less well documented, as they provide a gradation of symptoms among those afflicted by them. Of these, epilepsy may be mentioned, in which the chance of a sufferer among the brothers, sisters and children of an epileptic is 1 : 20 to 1 : 40; also gout, commoner in men than in women; probably essential hypertension; rheumatoid arthritis, in which there is a female excess; diabetes (mellitus) and schizophrenia.

To this multifactorial group belong, in general, the great mass of mental deficiencies. They are the result of large numbers of genes acting quantitatively, which endow their possessors with the lower range of mental qualities, and in the more extreme cases lead them to institutions. Here operates the converse of eugenics: for these people, more or less freed from a sense of responsibility, have larger families than normal. The situation is difficult to analyse, so that only in a few instances can it be determined that a major gene is paramount in the genetic control of the condition. This is restricted to cases in which some abnormal physical quality can be observed in addition to the mental one. An example of the kind is provided by the disease phenyl-ketoneuria. Here a single gene, recessive in effect, acts in two apparently distinct ways. It results in extreme mental deficiency, which can be distinguished from other similar situations because it is accompanied by the excretion of a substance (phenyl-pyruvic acid) that can be clearly identified in the urine.

This is the point to discuss a situation in which multifactorial inheritance and polymorphism may co-operate. They can do so in glaucoma, in which the pressure inside the eyeballs is raised because the fluid constantly secreted into them fails to escape as it should through the minute tube which exists for that purpose. At any rate in western Europe, this is the commonest cause of blindness in later life. The usual form of the disease is the 'open angle' one, in which no causative structural defect can be found in the eye.

There is considerable multifactorial variation in eye pressure even in normal people but, from middle age onwards, this may rise to the point at which it damages the retina: the layer at the back of the eye containing the light perceiving cells. Such damage is progressive, and takes place from the circumference inwards.

The defect is widespread in the population and extremely insidious, developing slowly over a period of years, so that the patient does not notice that his field of vision is contracting until the reduction becomes acute. It is then difficult to check, and the damage already done is irreparable. Treatment consists in the use of cortico-steroid drops and, when these are applied, the response sorts the population out into three groups, representing the two homozygotes and the heterozygotes of a pair of alleles, P^L and P^H which, in normal circumstances, merely contribute to the multifactorial variation in pressure already noticed. Their ratio is approximately $1 : 2 : 1$ among the offspring of the cross $P^L P^H \times P^L P^H$, and in the white population of the U.S.A. their percentage frequency has been established as $P^L P^L$ 66, $P^L P^H$ 29, $P^H P^H$ 5. The two latter groups, but especially $P^H P^H$, are those chiefly liable to the rise in pressure. The condition is to some extent environmental, and is probably affected by the blood pressure. It is more frequent in men than in women, though the genes concerned are transmitted equally by the two sexes.

It is necessary here to raise an important issue. All glaucoma cases should be warned that their children, in particular their sons, are in this matter at risk, and should be advised most strongly to have their eye pressure tested at least three times a year from, say, the age of forty onwards. In that way, the condition could be detected when treatment is still effective and little damage has been done. Such people would thus be saved from blindness or highly defective sight in later life.

The frequency of glaucoma is somewhat higher than normal in diabetics. Therefore when the one disease is detected, an examination should be carried out to make sure that the other is not developing also.

The unifactorial conditions, in which one gene plays a commanding part, need to be considered, however briefly, in several stages.

First may be mentioned the autosomal recessives, for these are often associated with the forms described under polymorphism. They appear here and there in a pedigree, depending upon the chance of marriage between two (unaffected) heterozygotes. It is clear that many of them must be maintained in the population owing to heterozygous advantage and so are, in reality, polymorphisms. A very few, however, may merely be the result of mutation. A small selection of such recessives may now be mentioned. Among these it will be interesting to think along the lines of human physiology: the way, that is to say, in which our body works.

There is hardly any growth in an adult mammal so that, except in a female carrying a foetus, only quite small amounts of nitrogen are needed; chiefly for repair and for the manufacture of certain secretions. The rest is excreted, mainly as urea. In the formation of that substance benzene is produced; but this is not easily eliminated from the body. However, one of the steps by which that can be done is attained by producing a substance known as phenylalanine, from which two pathways depart. One, by gaining a molecule of hydrogen and of oxygen, leads to the production of tyrosine and then on to melanin. This is a black or dark pigment related to indigo, the natural colouring matter of the woad plant, *Isatis tinctoria*; a substance figuring in the history books of our childhood as the somewhat inadequate clothing of the so-called 'Ancient Britons' who, we are informed, stained their bodies with it. Now the step from tyrosine to melanin is prevented by a gene, recessive in effect, which is therefore responsible for albinism. But it must have additional important consequences, since albinos have a shorter expectation of life than have normal people.

Albinism is variable in effect. The characteristic form has white hair; also pink eyes, due to the blood circulating in them not being hidden by pigment. In its less striking manifestation, the eyes are pale, watery blue; and the hair slightly yellow in childhood. All such people have defective vision, and tend to shun bright light.

A separate chemical route from phenylalanine is the one already mentioned leading to the disease phenyl-ketonuria (p. 157); while another gene, acting as a recessive, blocks a further stage along that line, so giving rise to the disease *alkaptonuria*. This tends to produce arthritis, and to cause blackening of the bones and cartilages.

We shall later need to use the information that certain genes are frequent in one or more races and effectively absent from others. Here we may usefully mention three instances of the kind. The production of a normal phosphate in the body occurs by forming, on the way, a type of sugar, 'xylulose'. At that point the action can be stopped, so that such sugar accumulates and passes into the urine,

F

sometimes leading to an incorrect diagnosis of diabetes. Such 'pentosuria' is recessive and almost confined to Jews: a people with two further characteristics that can well be mentioned here.

Two extreme forms of insanity, resulting also in early death, are the *amaurotic idiocies*. Both are simple recessives: one, starting in infancy, is especially a Jewish complaint; while from the other, arising a few years later in life, the Jews are largely exempt.

A famous example of a recessive disease also involving degeneration of the nervous system is *Friedreich's ataxia*. This at first affects the legs only: the patients sway as they stand; but later the body and then the head are involved; speech then becomes difficult, and the condition leads on to death. It generally starts during the second decade of life, and the cases appear to be sporadic, having normal parents. Children are hardly ever produced by the affected people. They would be normal except, sometimes, when the off-spring of cousin marriages.

Among many other diseases that are unifactorial and are auto-somal recessives, a few that claim attention are congenital dislocation of the hip, which leads to arthritis, also deaf-mutism. Here may especially be mentioned the recessive form of *retinitis pigmentosa*, which can be distinguished from several others because it alone is combined with deafness. The condition comprises a group of eye diseases to be contrasted with glaucoma, just described. The retinitis pigmentosas all start at any time from infancy to early adult life and involve a gradual degeneration of the retina. The first symptom is night blindness with, later, contraction of the visual field leading on to total loss of sight, as in glaucoma. From this they can, however, be distinguished, not only by their earlier onset, but by the spidery black spots visible at the back of the eye as the symptoms advance. These are due to pigment which has migrated from its original position behind the retina.

Some reference to sex-linked recessives will be found on pp. 45–8, taking haemophilia and colour blindness as examples of them. A number of other diseases are inherited in that way. Here it is worth while merely to point out that one of them is a further form of retinitis pigmentosa, but unaccompanied by deafness, the exis-tence of which has been noted. It is indeed an important fact that apparently similar conditions can be produced by distinct genes, not necessarily carried in the same chromosome, and sometimes expressing their effect in different ways as to dominance, reces-siveness and other qualities.

We can now turn to the situation in which a rare non-recessive disease or abnormality is controlled by a single major gene. This, not being masked in the heterozygote, will evidently be maintained in the population merely as a rarity; in fact by mutation subject to counterselection. Such a condition is generally described as 'dominant', but nearly always incorrectly. This section is not intended to include polymorphisms; yet it is probable that some conditions should be transferred to the polymorphic group, and this may well apply to the allergies.

It will be recalled that the term 'dominant' is applied only when a heterozygote and one of the homozygotes are similar in their effects. It is a state arrived at by selection in favour of a particular character, to make its heterozygous expression similar to the homozygous one if, and only if, this has an overall advantage. Yet when we examine a list of 'dominant' defects and diseases in man, we nearly always find that the rare homozygote has never even been seen; while in the few instances available for inspection, it is far more severe than the heterozygous state: that is to say, the two are not equivalent at all. Moreover, it will be noticed that the very serious diseases could never so evolve in the heterozygous state as to become recessives, since those who suffer from them leave few or no descendants on which selection could work. Thus such traits are 'heterozygous conditions', not dominants. A few of them out of several hundreds can be mentioned here.

Allergic diseases are due to a single controlling gene which so stimulates the body that it reacts in characteristic ways to a variety of environmental effects. In its presence, these may involve food allergies (often highly specific), urticaria, asthma, hay fever (both of the latter often stimulated by pollen), migraine and other symptoms. The strain imposed by a severe asthma attack may be sufficient to affect the heart, while severe heart disease can bring on asthma in an individual who has not previously experienced it. Thus, the association with the cardiac state may be effect or cause. The situation is much the more serious if the heart condition is the primary one.

In *brachydactyly* the usual (heterozygous) defect is that in which the fingers and toes are shortened to an extent that varies from one case to another. This is not too serious: after all, 'stump-fingered Mark' was, according to Iranaeus, able to write his Gospel, apparently dictated to him. Essentially it is the middle finger-bones that are reduced or malformed. At least two instances are known in which a pair of the heterozygotes married. Among the children of each, a sub-human creature appeared in which much of the skeleton was profoundly affected. Of course neither survived. It seems clear that these were the homozygotes.

Dwarfing can arise in a number of ways. There is, however, a distinct, extreme and well-known form, the 'achondroplastic dwarfs'.

They are readily identified, having short limbs but large heads, with a characteristic 'pug-like' facial aspect. They are heterozygous for a single controlling gene, so that half their children resemble them.

Though *Huntington's chorea* is an uncommon disease, it is frequently cited in discussing human heredity; and so it should be, for it has important features of a general kind. Like Friedreich's ataxia, it is due to a degeneration of the nervous system. Though occasionally appearing early in life, it usually develops between the ages of thirty and forty; but occasionally quite late, when the patients are over sixty. It involves the onset of involuntary movements, together with mental deterioration leading on to insanity. It can occur in either sex. Since it is a heterozygous condition, the gene is transmitted to half the children of anyone who develops the disorder. Two particularly sad aspects of the matter are to be noticed.

In the first place, since Huntington's chorea leads to incapacity and insanity, it is clear that normal people destined to develop it should not undertake family responsibilities. Yet, in general, the first signs do not appear until after the average age of marriage. Secondly, since its onset sometimes takes place in the elderly, or at any rate those over fifty, those who have had an affected parent live for most of their lives under the shadow of becoming insane, even if they have not inherited the condition and are in reality quite free from it. All that can be told to them at present is that they have equal chances of escaping the trait entirely; or of becoming incapable and insane in later life, and of passing on the gene to half their children, should they have any. An unfortunate feature of the disease is that in its earlier stages it increases sexual desire; consequently those afflicted by it tend to have more, not fewer, children than normal. A further point is to be noticed here. That is to say, a reasonably good human chromosome map, a quarter as good as already exists for a few other organisms, plants and animals, would make it possible to predict that the chances were, alternatively, small or great that a particular individual having a parent afflicted by Huntington's chorea had inherited the gene producing it.

Chromosome Abnormalities

Abnormalities in chromosome distribution and structure (pp. 108–13, 122) are responsible for certain human diseases. Such events have been selected against, and are consequently rare.

Abnormalities of individual chromosome *distribution* are a consequence of non-disjunction. That is, the two members of a pair pass together into the same reproductive cell and, on fertilization, the offspring will be trisomic in that respect.

The human chromosomes differ in size and shape. They are conventionally labelled from 1, the largest, down to 22, the smallest autosome, which is approximately as small as Y. X is about the mean size of, say, No. 10.

Non-disjunction of the sex chromosomes has already been mentioned. It occurs also in the autosomes, more often in some than others. Descriptions of diseases due to various types of trisomy are, indeed, numerous in the *Journal of Medical Genetics*.

Special attention must be drawn here to Down's syndrome, formerly called Mongolism or Mongolian Idiocy, an extreme form of mental deficiency, and with a characteristic physical appearance. It is generally due to the presence of an extra member of the twenty-first chromosome pair. There is an interesting environmental effect here. The chromosome abnormality increases in frequency with maternal age. In western Europe, when the mother is less than twenty-nine years old, one birth in 2,000 is a mongoloid; above the age of forty-five the frequency of that condition actually rises to one in 54 births. The victims are especially liable to leukaemia.*

Trisomics of chromosome 17, and within the group numbered 13 to 15, also give rise to severe mental retardation combined with various physical defects. These differ from one to another of such conditions, and from those that characterize mongolian idiots.

Two types of structural chromosome aberrations are especially relevant here. It has been mentioned that in deletion a piece of a chromosome breaks off and is lost. This has been extensively studied in experimental material (e.g. in the fly *Drosophila*). Some human patients with a form of leukaemia have one chromosome smaller than its partner. This seems to be the result of a deletion in Man. It has been established also that an insanity much resembling mongolian idiocy can be due to a translocation.

Cancer

It seems best to set cancer aside from other diseases, to be dealt with under a separate heading. For its manifestations are extremely diverse, cutting across the distinctions already made. Rightly enough, it is an illness that causes alarm and personal interest, so that readers

---

* In this fatal disease, the white blood corpuscles are greatly increased in number, owing to a cancer-like overproduction.

may well expect to find the subject treated as a whole, though briefly, in this book.

Cancer is the most ancient of diseases. We must think back for a moment to the remote epoch when living organisms began to grow large enough for their protoplasm to be subdivided into cells. One advantage of this is that the volume of cytoplasm in each is generally so small that it can be controlled by a single minute nucleus; one little larger than that required to house the chromosome set. There is, indeed, accumulating evidence for the view, put forward by myself in 1931, that the relatively large amount of cytoplasm in the egg of an animal and the ovule of a plant is too great to be controlled by its own nucleus. Consequently the course of its development is at first determined by the genes of its female parent; their effect being handed on up to the time when the nuclei of the offspring can take charge. That point is reached when the ratio of nuclear to cytoplasmic volume, at first much distorted, returns to normal as embryonic development proceeds.

When cells differentiate, they come to include inert material: in gland cells the product of their own activity; nerve and muscle cells acquire conducting and contractile fibres respectively; while skeletal cells manufacture supporting materials for the body. But inert substances hinder cell-division, which is therefore checked by differentiation. That complex situation must always have been subject to an occasional breakdown and failure: to cells losing or not acquiring their specialization, and so possessing the fatal quality of unlimited growth, as we see in the undifferentiated cells of the normal early placenta.* Consequently, they become cancerous or 'malignant'. Moreover, having nothing to hold them permanently together, some of these cells break away and float off in the body fluids, lymph and blood, to establish themselves elsewhere: that is to say, they form so-called 'metastases' which are additional cancers derived from that at the original site. In view of these facts, it is not surprising that we find cancers in all multicellular organisms: in plants and in lower and higher animals including, of course, Man.

A characteristic feature of this disease, then, is a rapid multiplication of unspecialized cells, so producing a growth which, pressing upon or eroding nerves, can be exceedingly painful. Such a mass of

* The placenta is the structure that in man, and most other mammals, connects the embryo with its mother, catering for the needs of embryonic respiration and nourishment. The early placental cells, derived from the embryo, make their maternal connection by eroding away the surface wall of the uterus. They occasionally fail to stop that activity, when they become actual cancers.

destructive cells has no proper blood supply, so that it is liable to decay and mortification.

What causes such cellular retrogression? There is both a genetic and an environmental element here, and it may sometimes be due to somatic mutation (p. 167). It may certainly be stimulated by continued irritation; but this leads to cancer in some organs, and in some parts of the body, more readily than in others. For instance, the mucous membrane of the nose is much subject to irritation, but it seldom becomes cancerous; it has been selectively adjusted to accept that stimulus. Also various human races and animal species differ greatly in the type and frequency of the cancers to which they are liable. Cancers of the colon and rectum are twenty times commoner in white men in the U.S.A., and ten times commoner in white women there, than in the corresponding sexes of negroes in Uganda, for which the evidence is good. Cancers are much more often a cause of death in domestic fowls (9 per cent) than in cattle (0·2 per cent).

The genetic component of cancer varies from those conditions in which the tendency to that disease is somewhat increased in certain families and gene-complexes to those in which it is unifactorial. Two instances of the latter type may be mentioned.

Multiple polypi of the colon are inherited as a single heterozygous abnormality, and lead invariably to cancer. There have been occasions, indeed, in which intelligent people, afflicted with the condition and seeing the fate of members of their family, have actually had their colon removed prophylactically before the malignant change had set in.

In xeroderma pigmentosum, the skin is abnormally sensitive to light. This becomes evident within a year of birth. Exposure to sunlight causes reddening of the face and hands, followed by the formation of severe freckles which do not disappear even when the skin is shielded from light. These later ulcerate. The skin appears abnormal in general; but only severely so on exposed areas, including the surface of the eye. In those regions cancers arise, from which the patient dies. The condition is nearly a recessive one, but heterozygotes tend to be heavily freckled. Evidently, two such individuals should not marry if they belong to families in which xeroderma pigmentosum has occurred.

Other unifactorial examples could be quoted. Usually, however, liability to cancer in general or in particular parts of the body, the female breast for example, is multifactorial, so increasing it in certain families. It is important to take this fact into consideration in medical diagnosis. For the possibility of this disease should especially be borne in mind in treating individuals from middle age onwards who belong to families in which several cancer cases have occurred, and of this the physician should be informed. It will be noticed also that there is a tendency for those who belong to blood group A of the ABO series to develop cancers of the stomach somewhat more often than other people; a fact of particular significance when treating stomach ulcers. The cancer frequency is also higher than normal in mongolian idiots and in triploids of one of the human chromosome pairs (No 21).

The genetic liability to cancer ensures that susceptibility to the disease can be affected by selection. This, however, could not influence what happens after the average age of reproduction,* though certain conditions that chiefly arise subsequently may once have done so earlier, having been pushed back in the life-cycle.

It seems that the human race had largely cured itself of cancer up to the end of mesolithic times: 2000 B.C., or a little earlier, in England. Few people lived beyond the age of forty until the more settled conditions of the neolithic period, when the new and relatively stable background of stock raising and agriculture allowed greater longevity. Mankind could then pass on into later life, when the onset of cancer had not been checked by selection. Consequently the more ordinary forms of the disease do not usually develop until middle age while, as I have in the past indicated, the connective-tissue tumours, which could not be selected against because of their rarity and danger, still occur in children and young people. We are here concerned with the fact that genes can control the time of onset of processes in the body, as shown by Huxley and myself in 1927.

One of the more hopeful ways along which the treatment of cancer may develop would be to induce the normal tissues to react against and destroy the malignant cells. There is some suggestion that this can occur; it is in part derived from studying those whose allergic reactions (p. 161) are particularly developed. Thus it has been found that while 12·9 per cent out of 294 normal people had allergic symptoms, these were present only in 3·2 per cent out of 1,185 cancer cases. Moreover, there is extensive evidence to show that in man and other animals that there is, in some instances at least, a reaction of the tissues against cancer cells; also that this is in some way related to the activity of the lymphocytes (a type of leucocyte, p. 140).

In Britain about 15 to 16 per cent of all deaths are due to cancer.

* We are about now reaching the stage when conscious selection might conceivably affect the frequency of cancer after the age of reproduction. One could imagine a tendency to avoid marriage into a family which had shown a marked incidence of the disease in later life.

Penetrating radiation greatly increases mutation in the body cells and in the germ-cells. The harmful effects of such 'somatic mutation' are liable to produce cancer of all types; but especially leukaemias, and lung cancers in those employed in radioactive mines. There is usually a long latent period before the growth appears: over twenty years in those exposed to radiation as an occupational hazard, and fourteen years or so following the more powerful dosage used in medical treatment; though skin cancers have been reported in a few weeks after radiation for cancer in the deeper tissues of the body. For this is used in therapy, based on the fact that cancer cells can be killed by a slightly smaller dosage of penetrating rays than are normal cells. The technique of using two sources with beams converging on the growth is therefore often an appropriate one. It is to be noticed also that the use of such radiation may be acceptable in the elderly, in view of the long latent period just mentioned before a radiation cancer is induced.

Perhaps a few words can usefully be said to those who have the slightest suspicion that they may be developing cancer: the discovery of a lump in the breast (this, of course, chiefly in women; about one breast cancer in 200 occurs in a man); any sign of blood in the urine or of rapidly developing difficulty to urinating; also in the other chief excretion from the body, if piles have been excluded; symptoms suggesting stomach ulcer. This rather frequently becomes malignant, while duodenal ulcers very rarely do so: a comforting fact not often realized by those who suffer from them. One must also add abdominal or thoracic pain or discomfort not dispelled reasonably soon by ordinary treatment; also, of course, the occurrence of an enlarging wart or skin ulcer, especially if coloured black or blackish, but even if small and unpigmented, when ulcerating and showing an inrolled edge. All such indications are of special importance in a family which has included several cases of cancer.

It should clearly be understood that many cancers can be treated successfully before the malignant cells have been scattered by the body fluids; afterwards, the situation, at least from a long-term view, is so far a hopeless one. Thus in this matter, early diagnosis and treatment are of the utmost importance. It is tragic how often the physician has to say, or wishes to say, to a cancer patient: 'Why did you not come for advice before?'

Smoking

The subject of cancer leads on to the danger of smoking tobacco. It must always be harmful and wholly unjustified to breathe an irritant substance such as tobacco smoke, but there seems no doubt that the hazard from cigarettes is significantly greater than that from cigars or pipes. Why this should be is not clear, but it is obvious enough that material differences are involved. In using cigarettes, paper is smoked in addition to the tobacco, which is itself of a different quality, differently cut, from other forms.

Lung cancer and bronchitis constitute the very real dangers of this habit. Those who seek to justify tobacco smoking have resorted to the suggestion that potential lung cancer patients have a greater desire to smoke tobacco than other people. There seems no basis whatever for this view; though there is, of course, a hereditary component in lung cancer, as in other forms of the disease.

The effects both of cigarette smoking and heredity have been analysed on several occasions, and the careful work of Tokuhata and Lilienfeld (1963) may be mentioned here. They used 270 lung cancer patients and their first-degree relatives compared with 270 controls and their first-degree relatives: 3,700 individuals in all. For comparative purposes, they attributed a risk of 1·0 to a non-smoker with no family history of lung cancer. For a non-smoking relative of a lung-cancer patient the risk becomes 3·96. It rises to 5·25 for cigarette smokers without affected relations, and to 13·64 for cigarette smokers with a relative who has had lung cancer. Evidently therefore, cigarette smoking and heredity can both contribute to producing a malignant growth in the lungs. Those who *inhale* tobacco smoke much increase these risks.

Lung-cancer patients sometimes try to excuse themselves by saying that they gave up smoking many years ago. In this they merely show their ignorance of the long latent period involved in the production of cancer by a given stimulus (p. 167). It should be added that tobacco smoking materially raises the incidence of bronchitis. Also, a heavy addiction to snuff is extremely dangerous: it is a powerful agent in producing the terrible condition of cancer of the mucous membrane of the nose.

8. Genetics and Human Societies

Polymorphism in Human Populations

Outbreeding leads to genetic variation, without which selection cannot operate so as to adapt organisms to changing conditions. Inbreeding, on the other hand, tends towards genetic uniformity, so preventing the breakdown of favourable adjustments to the environment. Indeed in its most extreme form, that of self-fertilization, it rapidly destroys genetic diversity. For we then have a breeding system in which all alleles already homozygous remain so, while half the heterozygotes become homozygous at each generation so that, by about the eighth, the result approximates to the effects of mutation; therefore at that point the process can go no further.

It has already been pointed out in Chapter 6 that since genetic variation is not related to individual needs, the probability of improvement is small indeed in organisms well fitted to their environment. Consequently, in such circumstances, they should vary as little as possible. But environments are not permanent things, and inbreeding does not supply the means of adjustment to new conditions, which is provided by outbreeding. Moreover, as any community becomes less well fitted to a changing habitat, the chances that random variation may throw up something useful become more likely.

Quite evidently, therefore, the perfect situation is that of the heterostyle-homostyle mechanism of plants (pp. 117–20), which allows passage back and forwards between outbreeding and inbreeding as required in any situation. Indeed, the plan is so satisfactory that it has been evolved very widely.* Yet the number of species that have been able to adopt it is not very great. What comparable devices are open to those that lack it, and to animals?

The genetic consequences of fluctuation in numbers provide a different means of adaptation to alternating good and poor conditions. We have seen (pp. 59–61) that in favourable surroundings a population will increase numerically, becoming more variable meanwhile, and so allowing new types of evolutionary adjustment;

* It has been found in eighteen orders of flowering plants.

while on the other hand, unfavourable changes in a habitat lead both to a reduction in numbers and to a more constant form. Moreover, such reactions are not restricted to plants, as in the heterostyle-homostyle device, but are of universal application, so that here too we find a method by which the breeding system can be adjusted to the environment.

There are, indeed, a number of other ways in which that end can be attained. These may best be thought of in regard to the subdivision of species into *races*, and especially so in regard to mankind. The distinctions between them are mainly multifactorial, producing a gene-complex characteristic of each. Its build-up depends on the nature and degree of the isolation, geographical or genetic, that characterizes the group. We see this beautifully illustrated in certain species that diverge gradually from one another over a long transect of country. Fertility may appear complete between neighbouring populations, yet the two ends of the series may be specifically distinct from one another. In birds, two North American warblers (*Phylloscopus plumbietarsus* and *P. viridanus*) are an example of that situation.

The problem of speciation has already been discussed briefly on pp. 73–7. There it was pointed out that even a slight degree of sexual isolation on crossing two races, shown by some abnormality in the heterogametic sex among the offspring, indicates that the first step to species formation has already been taken. It is finalized when sterility is complete, or effectively so. How does this relate to the races of mankind?

It cannot be doubted that such races exist, and a learned account of their characteristics and distinctions is given by Baker (1974). That they differ not only in their appearance, but in their anatomy and physiology, is a platitude that needs little discussion. Most of such features are multifactorial. They include height: for instance, Nordic and Japanese; and skull shape: that of the Eskimo is high, long and narrow, while that of the Laplander is remarkably short and wide. Also, of course, the shape of the mouth and nose are involved. There are, too, profound differences in the general shape of the body, in which an extreme departure from the average is shown by the female Bushmen. The great racial differences in skin colour are also multifactorial.

Yet the distinctions between the human races are believed to fall short of those indicating separate species. That is to say, crosses between them are held to be fully fertile, showing no disturbance even in the male offspring with their XY chromosome pair. That condition is liable to produce abnormal hybrid children because it is unbalanced by possessing the X chromosome of the race to which one parent belongs but not that of the other, unlike the XX sex which has one X from each.

Yet the non-specific status of the various human races is in fact a deduction that has never been fully tested; for none of the extreme crosses, being those most likely to reveal specific differences, if any, in mankind, have ever taken place: for instance between Eskimo and Australian aborigines, or between either of them and Bushmen. It is possible that boys might be abnormal, rare or absent among their children; we do not know. The distinction would also be particularly clear in F2, but that generation has never been raised from a human inter-racial cross.

This is due also to another important situation that affects the human breeding system: the incest taboo. It has been reinforced at an evolutionary level to produce an instinctive bias against incestuous unions. In general, the objection to such close inbreeding is partly due to its *social* consequences in the many societies in which the rights and status of an individual are precisely defined by kinship. In such circumstances, the offspring of incest become, as Bryan Wilson says (1975), a threat to order and stability; for their position in the hierarchy is unclear. The force of this was recognized also in the Christian Church. As Darlington (1969) points out, Pope Gregory prohibited marrying one's stepmother: an act of social, but of no biological, meaning.

The genetic aspect of the matter, though dim and wholly confused, has also long been current. Bede tells us that on the mission of Augustine to England, Pope Gregory instructed him to prohibit first-cousin marriages on the, wholly incorrect, view that such unions are childless.

Incest does not lead to infertility. Indeed if the community practising it continues to live in the unchanging situation to which it originally became adapted, it will now be realized that incest has exceptional advantages. But in a changing or a new environment, one to which those who emigrate to other lands are especially liable, inbreeding brings inevitable disaster, since an inbred population cannot adapt itself to fresh conditions.

The foolish and unworkable edicts of Justinian extended the prohibition of marriage far beyond brother and sister: indeed to the fifth degree. The Church accepted this ruling. Fortunately it was an impossible one, for by it the race would run into another danger; naturally enough when religion intrudes into a subject that it does not understand, and into which it has no business to enter. For there is indeed a balance of advantage and disadvantage between incest and outbreeding. In the one, the population may not be variable enough; in the other it may be too variable, since accurate genetic adjustments may be broken down by wide out-crossing. The objection, realized so clearly in India to the 'half-caste', the product of a marriage between a European and a native, is not only a social one, as those who have extensive experience of it would unhesitatingly endorse. Moreover, this applies to all such crosses from marriages between widely separate races; though each race may be excellent in itself. For the two will possess a gene-complex differently evolved and adjusted to different needs: two balanced systems that inevitably break down in a hybrid mixture.

We have here been thinking of situations that tend to maintain the human races and to exploit their existence as adjusted breeding groups. Another may usefully be mentioned at this point. That is to say, they may differ markedly in their scents. These can be highly objectionable to individuals of different racial types, though but little perceived within each. The point can be extended in a way that leads us again to consider the nature of variation, this time in relation to anthropology. It is said that about 10 per cent of Japanese produce a scent repulsive to the rest of the population. It has been plausibly suggested that the minority who possess this trait acquired it from a remote Ainu ancestor, belonging to the very distinct, hairy and strong-smelling people in the extreme north of the Japanese islands.

It is to be noticed how much more variable are some races than others: adjusted, that is to say, on the one hand by overall segregation and on the other by means of an average (pp. 31–2). We find it easy to recognize similarities or distinctions in populations of a kind that we know. A Highland farmer, taken to an art gallery in Edinburgh and shown a picture of a shepherd and his flock, burst into laughter, saying: 'But they are all the same sheep, they have all been painted from one animal.' Doubtless Japanese appear more variable to their own countrymen than they do to Englishmen, but there is reality here. An unbiased observer from another planet would find the Japanese much the less diversified: far more similar than the English in height, skin colour and facial shape; all with straight, black hair and dark eyes.

Scent production in the Japanese, just mentioned, is clearly a polymorphism within a racial feature that seems generally multifactorial. Indeed many obvious racial characteristics are polymorphic, or show combinations of that condition with one in which the variation is 'continuous'. Furthermore, the gene-complex affects the expression of the polymorphic phases. Thus, red hair is a simple recessive, often with a trace of heterozygous effect, while a number of genes act cumulatively to produce the shades from light brown to black. Similarly with eye colour; blue (turning to grey) is a single recessive, while the diversity of brown eyes is due to genetic modifiers.

It is convenient to introduce polymorphic racial characteristics by means of a feature fundamental to man and to mammals generally, the formation of lactase. This is an enzyme that breaks down milk sugar. Owing to its absence, most adults in the world are unable to drink fresh milk. Naturally, all infants possess lactase, which begins to be formed near the end of foetal life and, upon reaching a peak soon after birth, is progressively reduced in amount and ceases to be manufactured shortly after weaning ends. However, a gene dominant in effect can continue its production throughout life. Though generally absent, this is found in about 90 per cent of northern Europeans and in about 80 per cent of two nomadic pastoral tribes in Africa, the Fulani of Nigeria and the Tussi of Uganda. Both are dependent upon milk, and in them there has evidently been selection for the gene extending lactase production.

We can now think in more general terms of the light thrown upon the human races by polymorphism. Its importance in this connection is very great, as we should expect of that type of variation, since it is concerned in holding different forms together. From that point of view the human blood groups are of outstanding significance in anthropology. It is indeed extraordinary that they were so long held to be of neutral survival value, even though it was known that interactions between them could be fatal! The inconsistencies of such a view were indicated by myself in 1940, while soon afterwards (1942) I showed that the respective blood-group series must each be polymorphic, with all that this implies.

When reviewing the matter in 1957 I was dealing with the blood groups, but my remarks were applicable to all polymorphisms. Though often quoted, what I said may be repeated here. The passage runs as follows:

> By a curious inversion of logical thought, it was held that their occurrence [that of the blood groups] in distinct and characteristic proportions in the different races of mankind was especially important because the variation involved was selectively neutral. Precisely the contrary is true. The fact that the genes concerned are balanced by selection at optimum frequencies, which differ from race to race, is the one which gives them significance as a criterion of relationship. It does so because in these circumstances their proportions are influenced by the average genotype of the population in which they occur.

It is in analysing the physiological features of the different human races that genetics has so far contributed most. Many of these are polymorphic, maintaining distinct character frequencies in each. For instance, non-tasters of phenyl-thio-urea (p. 149) comprise about 20 to 30 per cent of western Europeans, but 10 per cent or less of American Indians; down to 2 per cent in some of the tribes.

We can now briefly consider the important contributions of the blood groups to the study of human races. It is indeed fortunate that an up-to-date survey of the subject on a vast scale has recently been published (by Mourant and his colleagues, 1976). It will probably be best at the outset to give a short account of the basic features of the blood groups in different continents. It will then be possible to think of their distribution in certain races. These can be chosen for their special interest and, in some instances, for the light that blood grouping throws upon history. Here and there, the matter can also be illuminated by brief reference to some of the other polymorphisms mentioned in Chapter 7.

Here we are principally limited to the ABO and Rhesus groups, being those already discussed in a little detail (pp. 144–8). They are indeed the most important, and by far the best known; but the numerous other blood groups, as well as all polymorphic features, provide additional, and sometimes critical, information also.

The southern English values of the main ABO groups, in percentages, are O, 43·5; A, 44·5; B, 8·6; AB, 3·2. They do not differ greatly from this in the chief countries of western Europe; this is rather a high proportion of A compared with the rest of the world. The A2 subdivision of A, to which about 10 per cent of southern English belong, is almost restricted to Europe and is, in consequence, the most outstanding feature of the ABO series there. It was thought that it occurs also, and at about the same frequency, in negro blood; but this has now been corrected. The African and European A2 groups are distinct.

The value of B rises in eastern Europe to exceed 10 per cent. Except for eastern Germany, the boundary of the area where it does so accords fairly closely with the western limit of Slavonic languages. The increase in B continues south-eastwards until it reaches its maximum for the world, 37·2, among the Hindus of northern India. This is nearly retained up to Java and Borneo, but diminishes further eastwards. However, it is noteworthy that group B is absent among Australian aborigines.

The Rhesus-positive supergene $D^A C^A E^B$ is the commonest of its group in southern England (40·8 per cent); while $DC^B E^B$, being that for the definitive Rhesus-negatives, is very similar at 38·9. This is the frequency per chromosome. Being recessive in effect, it ensures that about 15·1 per cent of the population are Rhesus-negatives (p. 147); a value that is maintained throughout western Europe, though a little less round the Mediterranean.

To this statement the Basques provide a striking exception. They represent a survival of the ancient (late-palaeolithic) inhabitants of Europe, as indicated by their skeletal features, the region (of the Pyrenees) in which they live: driven north-east in Spain and south-west in France, while alone among the people of western Europe, they have retained a non-Indo-European language.*

The Basque blood groups support this view. In them group B is so rare that it may well be fundamentally absent, the few individuals possessing it having obtained it from hybrid French or Spanish unions. Their Rhesus-negative supergene is remarkably common, at a chromosome frequency of 55 per cent. Indeed the normal west European value may well have been arrived at as a mixing during the last few thousand years of a predominantly Rhesus-negative population, of which the Basques are a survival, with a predominantly Rhesus-positive one such as occurs widely in Asia, and may have been imported to the West with the neolithic incursions from the East, responsible for agriculture and stock raising. That culture began to infiltrate Southern England from about 2200 B.C.

On the whole, the most striking serological feature of the native races of Africa is the great preponderance of the $D^A C^B E^B$ supergene, where it occurs at about 60 per cent. This is to be compared with 2 per cent in Europe, and less than 15 per cent in the rest of the world.

The three great racial types of America can be rather clearly distinguished by their ABO blood groups. From Mexico southwards, group O is so overwhelmingly common that it may have been the only one present prior to the arrival of the Europeans. Among North American Indians, group O is also strikingly frequent; but A also occurs, though in highly variable amounts from one tribe to another; while B seems genuinely absent. Non-secretors of the ABO antigens are extremely rare among these peoples. Both in North American Indians and in South America the $D^A C^B E^A$ supergene reaches the highest value attained anywhere in the world. The third group, the Eskimo, are polymorphic for all three of the ABO alleles; their genes for groups A and B occupy respectively about 40 to 45 and 8 per cent of the total. In their Rhesus genes they resemble the North American Indians, yet nothing could be more different than the reaction of these two peoples to tasting phenyl-thio-urea. In the

---

* The language of the Lapps, also non-Indo-European, is almost certainly derived secondarily from Finnish.

G

one, non-tasters reach the lowest (under 10 per cent) and in the other, the Eskimos, the highest (40 per cent) levels known.

We can now think of the blood groups in certain other regions and races.

The situation in rural Ulster was examined by Hart (1944), who found that blood donors with English names have English blood-group values, of the Midlands and the South, though it is more than three hundred years since their ancestors emigrated thence. The rest of the Ulster population has blood groups of the Lowland Scottish type. These differ in the remainder of Ireland, where they show Highland and Icelandic affinities, partly, no doubt, through Viking incursions.

The intrusion into Europe of peoples from the Near and Far East forms a fascinating story: one on which the blood groups throw important light.

Attention has already been drawn by Mourant to a suggestive aspect of ABO in this respect: the existence of 'concentric tongues of high B and low O frequencies based on the northern end of the Caspian Sea and sweeping across southern Russia and Poland into eastern Germany'. These, as he says, surely indicate the routes taken by immigrating hordes from the East making their way into central Europe.

It is now accepted that the Gypsies originated in India. This is clear from their physical attributes and on linguistic grounds; also from their blood groups. It seems certain that the Gypsies made their way from Eastern into Western Europe during the period from A.D. 1300 to 1500, and that they had probably reached the Near East about the year 1000. We do not know from what part or parts of India they came: they may have originated from more than one region; nor do we know the date of their emigration or the reason for it. However, Mourant plausibly suggests that their departure may have been due to the Moslem invasions.

The Gypsies have of course often crossed with local populations to produce hybrid offspring who prove particularly unsatisfactory (p. 172). However, the pure-bred type has been preserved to a remarkable degree. It constitutes a race scattered but consistent, retaining clear indications of its Indian ancestry. The frequency of it B blood group is far higher than that of the peoples into which it has intruded; indeed it is approximately of the Indian value, exceeding 30 per cent. This is to be compared with its cline from 5 per cent in the west of Europe to 12 to 15 per cent in the south-east. That result is obtained, as in India, by reducing the group O frequency, to under 30 per cent; instead of over 40, which is its normal European level. The Rhesus series provides independent, and therefore important, testimony of a similar kind. Thus the Rhesus-negatives comprise about 16 to 20 per cent of the normal populations in Western Europe, 12 to 15 in the Eastern part of the continent. They are rarer among the Gypsies, about 4 to 8 per cent. That is to say, they take values within the Indian range of 1·5 to 10 per cent.

It is necessary to take up, though briefly, a racial problem of particular difficulty, that of the Jews. Though scattered so widely, they differ in their physical and mental traits from the Gentiles among whom they dwell. Yet it must be noticed that most of the Jews in Europe and North America are from the Mediterranean region, and basically Semitic. Their most consistent features are a fairly high frequency of blood group B (15 per cent or so) and a low proportion of Rhesus-negatives. Moreover, they have other characteristic qualities: for instance, the disease pentosuria is almost confined to them, as is one of the two forms of amauritic idiocy, while from the other they are largely exempt.

On the other hand, conversion has been widely accepted by the Jews; and this must introduce much heterogeneity among them as a whole. One finds, indeed, extreme examples of this in the Chazare in Russia, Jewish but not Semitic, and in the Chinese Jews with Mongolian qualities due to intermarriage. There remains also the black element in the population of the Levant: one sees wholly negroid Mohammedans in Jericho, maintained even now, just as occasional black Jews are to be found in Israel.

Indeed the contrast between the Jews and the Gypsies is extreme. It is particularly revealing when we take into account ancient 'built-in' genetic adjustments compared with those that are labile (pp. 180–1).

It is approrpriate to mention the Samaritans here; an interesting relic rapidly declining in numbers but retaining their identity almost undiluted. They are Israelites but not Jews, the only ascertainable remnant of the Northern Kingdom. They retain traditions to indicate their respective tribes, of which a few have been grouped for their ABO frequencies (Fig. 8.1):

	O	A	B	AB	totals
Ephraim	54	10	4	—	68
Manasse	36	17	13	4	70

Fig. 8.1. Samaritan ABO blood groups.

Even on these small numbers, the use of probability mathematics shows that there is a real difference between the ABO blood groups of Ephraim and Manasse. Clearly the physical characteristics, serology and traditions of the Samaritans should be examined in detail before their identity is lost.

Turning to another subject, the adventure of Heyerdahl (1952) in setting out westwards across the Pacific on a raft of balsa logs has invested the origin of the Polynesians with romance and popular appeal. For this was the problem he was attempting to solve, believing it was thus that this people colonized the Pacific islands as far as Tonga, with an extension to New Zealand to include the Maoris.

The Polynesians are brown-skinned, with straight, black hair. Thus they differ greatly in appearance, and indeed in culture, from the Papuo-Melanesians, who are intensely black, with fuzzy hair; the latter group extends eastwards as far as Fiji. Indeed, traces of such a dark, curly-haired people allied to the African negroes are found in Indonesia, the Philippines and the Andaman Islands, and even in India.

Are the Polynesians, as formerly thought, from Asia or, to some extent at least, from America? Heyerdahl holds the latter view, and thinks that they are derived from two distinct migrations from the New World: one, of a fair-haired race which possibly reached South America from the Canary Islands and later sailed westwards across the Pacific on balsa rafts; the other of American Indians from the west coast of Canada: the Kwatiutl tribe, which still survives, is regarded as its source.

It seems that while there is truth in both these propositions, their supporters have exaggerated the importance of them. First, we may summarize the American affinities of the Polynesians. The gene for group B of their ABO series takes so low a value that originally it may well have been absent, as throughout all American natives (except for the Eskimo). This is the very reverse of the Asiatic situation. It is true that the Australian aborigines also lack B, but there is no evidence to suggest that this most primitive people have contributed to the genetics of the races surrounding them today. There are other features associating the Polynesians particularly with North America. We may cite their high frequency of group A of their ABO series; also of the Rhesus supergene $D^A C^B E^A$, which reaches 54 per cent. Indeed they and the tribes of British Columbia are unique in possessing this as their most frequent Rhesus combination, actually exceeding $D^A C^A E^B$. The Central and South American affinities of the Polynesians are also indicated by their characteristic crop, the sweet potato, *Ipomoea batatas*, which they must have brought thence; while their cotton is American in type. (So also are two of their other blood groups, MN and P, not discussed in this book.)

On the other hand, the Polynesians must have obtained the pig from Asia; and the high value of non-secretors of the ABO antigens, 31 per cent, is clearly Asiatic, for that condition is virtually absent from American Indians. Moreover, Polynesian archaeology also suggests an Asiatic source, while their language is Melanesian in general type and has no connection at all with anything in America.

Surveying this matter in general, we can say that the Polynesians seem to be derived partly from South America, but that emigrations from the west coast of Canada have been more important in their evolution. Also that they have certainly received Asiatic contributions to their culture and genetics.

Blood groups occur in other mammals besides man; and it is tempting to speculate on the primitive blood grouping of mankind. There is perhaps little ground for doing so, since similar serological adjustments may have arisen independently. Perhaps, however, it may not be unreasonable to suggest that the absence of group B of the ABO series is a primitive feature. It does not occur among Australian aborigines, whose civilization is palaeolithic, or among those of South America: these being the two areas of the world where the most ancient types of living organisms are congregated. Moreover, group B is absent from the Basques, who antedate the normal European types. On the other hand, it is rather common among the certainly primitive Bushmen. It is, indeed, at a more recent level that the blood groups make their most significant contribution to anthropology.

It would not be possible to read even the short account of human races given here without being impressed by the stability of their genetic control; and we have touched only the fringe of the matter. In this connection, we may think of the English population in Ulster, retaining for centuries the blood-group values it had acquired before its arrival; of the persistence of the Indian frequency of blood group B among the Gypsies of western Europe after a thousand years of wandering; and of the fact that the Polynesians retain the unique value of their $D^A C^B E^A$ supergene already acquired in their far-off homeland of British Columbia. A further dimension is added to the matter by Mourant in quoting unpublished work by Messerlin in north-west Africa. He had examined the blood groups of the Aït Slimane of the Great Atlas: a Moslem tribe having a tradition that they were Jews who became converted to Islam; and indeed they retain many Jewish customs. Their blood groups carry us back one stage further and show that prior to their Jewish past they were Berbers, like their neighbours; for they preserve unaltered their Berber blood groups today.

Such extreme stability is to be contrasted with the great genetic variability upon which selection can act to produce rapid evolution. Thus the occurrence of sickle-cell anaemia, due to heterozygotes developed to withstand malignant tertian malaria (pp. 151–2), is so easily adjusted that in Africa it quickly declines in response to public-health measures designed to eliminate the disease. Among much else of the kind, our thoughts turn also to the striking poly-morphisms of industrial melanism in which, for instance, the gene for the dominant black form of the moth *Biston betularia* spread around Manchester from extreme rarity, mutation level in fact, to affect 98 per cent of the population in less than half a century.

How can these things be?

We turn back to Chapter 3 of this book (pp. 155–6), where it is pointed out that the Mendelian system is capable of ensuring great heritable *variability*, on which selection can act to promote rapid evolution, and great heritable *stability* to retain those qualities, and combinations of qualities, of advantage to the organism; and we have now seen both aspects of that apparent paradox working harmoniously.

It is mainly with stability that we are concerned in assessing the genetics of the human races, and for the reason that they each construct, independently, gene-complexes adjusted to their needs: a balance broken down by intercrossing them. In these races, there-fore, are operating sets of genes that must strongly resist minor genetic and evolutionary changes. These, with their appropriate genetic build-up, have been named *palaeogenes* by McWhirter (1967). On the other hand, relatively newly used major genes, and those that must be adapted to fundamental situations recently arisen, will not have acquired such accurately adjusted genetic settings, and these McWhirter named *neogenes*.

The blood groups are palaeogenic. The polymorphism of in-dustrial melanism, for example, is neogenic; although evidently requiring the evolution of an associated gene-complex to produce heterozygous advantage and other physiological adaptations. For such melanism is recent, and has arisen in response to recent con-ditions. Consequently the frequencies of its phases can rapidly follow relevant environmental changes such as the introduction of smoke-less zones (p. 75.); the very epitome of what the blood groups cannot do and, indeed, are protected from doing.

We find, then, a striking persistence of blood-group types among the Jews in Canada who are of Mediterranean origin; while in a wider context conversion has introduced diversity among those holding Jewish beliefs: diversity of a kind not to be encountered among the Gypsies; and this apart from the outstanding mental qualities to which the Jews can attain.

Even in discussing at a superficial level two of the blood groups in Chapter 7, attention was drawn to the danger of inappropriate marriages within each: giving rise, for instance, to a group A foetus in a group O mother (p. 149), or when a Rhesus-positive husband has a Rhesus-negative wife (p. 147). Types producing such 'mistakes' could not persist in the population unless they carried counterbalancing advantages with them. The reason for their existence is that a polymorphism is involved, so holding populations and races together.

There is too much tendency to evaluate the blood groups separately, without considering that the genes controlling them must interact with one another and with the gene-complex as a whole. Not only is it clear on the evidence of general genetics that such interactions must occur, but examples of them are well known in Man. We have here one of the reasons for the long-term stability of the blood groups in different races.

Language and Intelligence

The basic steps that led to human evolution are indicated on pp. 76–7. Two of its essential subsequent features must now be mentioned: the invention of speech; and selection for high intelligence, itself, as already pointed out, the key to the whole process. First, in relation to speech, we must think of the phonetic aspect of language, a subject brilliantly discussed by Darlington (1947, 1969) and by Brosnahan (1961, 1962).

Darlington points out that in addition to the brain, speech has required much co-adapted evolution involving the larynx, palate, tongue, teeth and lips, together with their nervous control. These have developed sometimes divergently, sometimes in parallel, in the races of mankind. At least a hundred basically distinct language *types* are known, with thirty times as many mutually unintelligible languages. Their evolution has been responsible for the major distinction between man and other animals; such physical and mental adjustments must have required a great period of time. The evolution of speech is ancient and slow, and that of language is recent and rapid: so rapid that in England at the end of the Middle Ages, as a result of the fusion between Anglo-Saxon, Latin and French, the pace was becoming intolerable. When Caxton developed printing, and incidentally stabilized the language thereby, he was bemused by what was taking place. 'Am I', said he, 'to print the language as I heard it when I was a boy or as I hear it, profoundly altered, around me now?' And today we find, not only in time but also of course in space, differing importance attached to vowels, consonants and intonation. Thus the latter is used more frequently to express the meaning of a word in some languages, as Chinese, than in others, those of Europe; though even there it can take a part, showing the possibility of its development (Latin: pōpulus = a poplar tree, pŏpulus = the people). Moreover, as Darlington says, every race 'has a genetically different sound-producing apparatus and all prefer to use the sounds, and combinations of sounds, that come most easily to them'. Therefore languages change and evolve with the people who speak them, and such distinctions of speech are genetic. Of that fact, an illustration has been provided by Darlington (1947), who has shown that there is a striking agreement between those who use the ð and þ sounds* and the races with 64·5 per cent or more of the gene *G* of the ABO blood-group series, except for the Italian peninsula and the islands between it and Spain.

Turning to intelligence, this of course differs from one individual to another, unless they be identical twins. In order to compare intelligence, it must be measured. In this matter, we are concerned to assess the power to comprehend and to reason: *cognitive ability*, it is called. It is important to distinguish this from acquired knowledge of which, owing to the opportunities of education and experience, the fool may possess more than the genius. That is what is attempted in intelligence tests. These are of very dissimilar kinds, each of which has been modified and improved. If the cognitive ability of average children in the more advanced European countries be rated as 100, the percentage of that value is known as the *intelligence quotient* (I.Q.). It has been found meaningful to rate an I.Q. of 65 to 70 as 'mentally defective' and of 140 and upwards as 'gifted'.

This is not the place to discuss the details of intelligence tests. They can be verbal and non-verbal, but it is most important to notice that the results obtained by means of extremely different types of tests agree very well. A step of outstanding importance was taken when A. H. Wingfield showed that the nearer the relationship between individuals may be, the more closely, on the average, they resemble one another in performing intelligence tests, though here and there quite outstanding ability becomes apparent. The genetic

---

* ð being the *th* as in 'then', and þ being the *th* as in 'thin'.

basis of intelligence is also well demonstrated by the fact that the I.Q. of identical twins reared apart is much more alike than that of unrelated people reared together.

A further type of evidence is obtained from a comparison between the intelligence of relations. Thus a group of 3,400 children of school age were given mental tests and classified as 'bright' (I.Q. of 113 and over), 'average' (I.Q. of 91 to 113) and 'dull' (I.Q. of less than 91). Three classes were then selected: the brightest 4 per cent, the central 4 per cent and the dullest 8 per cent. The brothers and sisters of school age children of these three groups were also tested. It then appeared that 62·3 per cent of the brothers and sisters of the brightest children were bright, and only 6·6 per cent were dull; while only 3·7 per cent of the brothers and sisters of the dullest children were bright, but 56·3 per cent of them were dull.

Tests along these and other lines have shown many times that much of the variation in intelligence is genetic. Thus, selection can operate effectively to improve human mental qualities.

Since intelligence varies genetically from one individual to another, so on the average it is bound to differ from one race to another: a fact established by many authorities in a number of countries. It would appear almost unbelievable, but true, that since the passing of the recent Race Relations Act it has become illegal in Britain to publish facts demonstrating intellectual inequality in the human races. We were of course familiar with a similar situation in Nazi Germany, where ascertained facts proving the ineptitude of current political theories were also legally suppressed. This, however, appears to be the first instance of the kind in Britain. Consequently the facts illuminating this important subject cannot at present be included in this book. They can, of course, be obtained from numerous works published in America.

Finally, attention must be drawn to another aspect of intellectual variability and its genetic basis. That is to say, the need in society for hereditary social distinctions, but of such a kind that it is possible to rise to the higher levels or sink to the lower ones. I was once present at a political meeting in England at which the speaker, voicing the attitude of his party, said: 'We are not so foolish as to suggest that all men are equal, of course they are not; but we do hold that all men should be given equal opportunities.' The proposition is not a practicable one but, if it were, one aspect of it seemed to have escaped the speaker's attention. Since ability is inherited, such a programme would effectively generate or preserve class distinctions. That result, desirable as it is (Darlington, 1969), has in the past been achieved more realistically by giving better opportunities to those best fitted to use them.

REFERENCES

Baker, J. R. (1974). *Race*; Oxford Press.

Bradshaw, A. D., McNeilly, T. S. and Gregory, R. P. (1965). 'Industrialization, Evolution and the Development of Heavy-metal Tolerance in Plants'; *5th Symposium Brit. Ecological Society*, 327–43; Blackwell, Oxford.

Brosnahan, L. F. (1962). 'The Influence of Genetics on Language'; *Discovery*, **23**, 400–3.

Cain, A. J. and Sheppard, P. M. (1950). Selection in the Poly-morphic Land Snail *Cepaea nemoralis*; *Heredity*, **4**, 275–94.

Cavalli-Sforza, L. L. and Bodmer, W. F. (1971). *The Genetics of Human Populations*; Freedman, San Francisco.

Clarke, C. A. (1964, 2nd edn.). *Genetics for the Clinician*; Blackwell, Oxford.

Clarke, C. A. and Sheppard, P. M. (1960). 'The Genetics of *Papilio dardanus*'; *Genetics*, **45**, 439–57.

Creed, E. R. (1974). 'Two-spot Ladybirds as Indicators of Intense Local Air Pollution'; *Nature*, **249**, 390–2.

Crew, F. A. (1925). *Animal Genetics*; Oliver and Boyd, Edinburgh.

Darlington, C. D. (1937, 2nd edn.). *Recent Advances in Cytology*; Churchill, London.

(1947). 'The Genetic Component of Language'; *Heredity*, **1**, 269–86.

(1963, 2nd edn.). *Chromosome Botany*; Allen and Unwin, London.

(1969). *The Evolution of Man and Society*; Simon and Schuster, New York.

Dobzhansky, T. (1950). 'Genetics of Natural Populations', XIX; *Genetics*, **35**, 288–302.

Dobzhansky, T. and Pavlovsky, O. (1966). 'Spontaneous Origin of an Incipient Species in the *Drosophila paulistorum* Complex'; *Proc. Nat. Acad. Sci. Wash.*, **55**, 727–33.

Fisher, R. A. (1936). 'Has Mendel's Work been Rediscovered?'; *Ann. Sci.*, **1**, 115–37.

Ford, E. B. (1931). *Mendelism and Evolution*; Methuen, London.

(1940). 'Polymorphism and Taxonomy', in *The New Systematics* (ed. Julian Huxley); Oxford Press.

(1942, 1st edn.). *Genetics for Medical Students*; Methuen, London.

(1955). 'A Uniform Notation for the Human Blood Groups'; *Heredity*, **9**, 135–42.

(1971, 6th reprint). *Butterflies*; Collins, London.

(1973, 7th edn.). *Genetics for Medical Students*; Methuen, London.

(1975, 4th edn.). *Ecological Genetics*; Chapman and Hall, London.

(1976, reprint). *Moths*; Collins, London.

and Huxley, J. S. (1927). 'Mendelian Genes and Rates of Development in *Gammarus chevreuxi*'; *Brit. J. Exp. Biol.*, **5**, 112–34.

Ford, H. D. and Ford, E. B. (1930). 'Fluctuation in Numbers and its Influence on Variation in *Melitaea aurinia*'; *Trans. Roy. Ent. Soc. Lond.*, **78**, 345–51.

Fujino, K. and Kang, T. (1968). 'Transferrin Groups in Tunas'; *Genetics*, **59**, 79–91.

Handford, P. T. (1973, *a* and *b*). 'Patterns of Variation in a Number of Genetic Systems in *Maniola jurtina*'; *Proc. Roy. Soc. Lond.*, B, **183**, 265–84, 285–300.

Hart, E. W. (1944). 'An Analysis of Blood-group Composition in a Population in Northern Ireland'; *Ann. Eugen.*, **12**, 89–101.

Heyerdahl, T. (1952). *American Indians in the Pacific*; Allen and Unwin, London.

Huxley, J. S. and Ford, E. B. (1925). 'Rate Genes'; *Nature*, **116**, 861–3.

Ives, P. T. (1950). 'The importance of Mutation-rate Genes in Evolution'; *Evolution*, **4**, 236–52.

Kettlerell, H. B. D. (1973). *The Evolution of Melanism*; Oxford.

Mourant, A. E. *et al.* (1976, 2nd edn.). *The Distribution of the Human Blood Groups*; Blackwell, Oxford.

Nicholls, E. M. (1969). 'The Genetics of Red Hair'; *Human Hered.*, **19**, 36–42.

Perrine, R. P. and others (1972). 'Benign Sickle-cell Anaemia'; *Lancet*, **2**, 1163–7.

Race, R. R. and Sanger, R. (1969, 5th edn.). *Blood Groups in Man*; Blackwell, Oxford.

(1976). Blood-group Polymorphism; *Transfusion and Immunology*.

Rothschild, the Hon. M. (1972). 'Some observations on the relationship between Plants, Toxic Insects and Birds', in *Phytochemical Ecology* (ed. J. B. Harborne), pp. 1–12; Academic Press, London.

(1971). 'Speculations about Mimicry with Henry Ford', pp. 202–23 in *Ecological Genetics and Evolution* (editor E. R. Creed); Blackwell, Oxford.

(1972). 'Secondary plant substances and warning coloration in insects', pp. 59–83 in *Symposium No. 6 of the Royal Entomological Society of London*; Blackwell, Oxford.

and Clay, T. (1952). *Fleas, Flukes and Cuckoos*; Collins, London.

and Reichstein, T. (1976. 'Some problems associated with the storage of cardiac glucosides by insects', *Symposium on Secondary*

Metabolism and Coevolution (editors Luckner, M.; Mothes, K.; Nover, L.), *Nova Acta Leopoldina, Supplement* 7, Deutsche Akadamie der Naturforscher Leopoldina, Halle.

Rowse, A. L. (1977). *Homosexuals in History. A Study of Ambivalence in Society, Literature and the Arts*; Weidenfeld and Nicholson, London.

Stebbins, G. L. (1963). *Variation and Evolution in Plants*; Columbia, New York.

Tokuhata, G. K. and Lilienfeld, A. H. (1963). 'Familial Aggregation of Lung Cancer in Humans'; *J. Nat. Cancer Inst.*, **30**, 289–312.

Vogel, F. and Chakravartti, M. R. (1966). 'ABO Blood Groups and Smallpox'; *Humangenetik*, **3**, 166–80.

Wahrman, J. and others (1969). 'Mole Rat *Spalax*: Evolutionary Significance of Chromosome Variation'; *Science*, **164**, 82–4.

Westerman, J. M. and Parsons, P. A. (1973). 'Variations in Genetic Architecture at Different Doses of Gamma Radiation Measured by Longevity in *Drosophila melanogaster*'; *Canadian J. Genet, Cytol.*, **15**, 289–98.

Wilson, B. R. (1975). *Education, Equality and Society*, pp. 9–38; Allen and Unwin, London.

Wylie, A. P. (1954). 'Chromosomes of Garden Roses'; *Amer. Rose Annual, 1954*, 36–66.

BIBLIOGRAPHY OF BOOKS BY E. B. FORD

Mendelism and Evolution, 1931, 8th edn. 1965, Methuen and Co., London (also in Science paperback edn.; 2nd edn. of Spanish trans. 1968, Molins de Rey, Barcelona).

The Study of Heredity, 1938, Home University Library; 2nd edn. 1950, Oxford University Press.

Genetics for Medical Students, 1942, 7th edn. 1973, Chapman and Hall, London (Italian trans. published as *Genetica*, 1948, Longanesi, Milan.

Butterflies, 1945, 4th edn., after three reprints of the 3rd edn., 1977, New Naturalist Series, Collins, London (also in Fontana paperback edn., 1975).

British Butterflies, 1951, King Penguin Books, Harmondsworth.

Moths, 1955, reprint of 3rd edn. 1976, New Naturalist Series, Collins, London.

Ecological Genetics, 1964, 4th edn., enlarged, 1975, Chapman and Hall, London (Polish trans. 1967, Warszawa; French trans. 1972, Gauthier-Villars, Paris; Italian trans. in press for 1978, Zanichelli, Rome).

Genetic Polymorphism, 1965, Faber and Faber, London.

Evolution Studied by Observation and Experiment, 1973; Oxford Biology Readers, Oxford University Press (2nd edn., enlarged, in press).

Genetics and Adaptation, 1976, Institute of Biology Studies, Edward Arnold, London (also in paperback edn.).

(with G. D. H. Carpenter). *Mimicry*, 1933, Methuen and Co., London (Spanish trans. 1949, Acme Agency, Buenos Aires).

GLOSSARY

Aestivation A quiescent state that enables some animals to survive the heat of summer (compare *Hibernation*).

Agouti A brownish hair colour due to narrow alternating bands of black and yellow.

Alleles (*Allelomorphs*) Genes occupying identical positions in homologous chromosomes. They therefore separate from one another during meiosis. Such genes control in differing degrees the same set of characters.

Allergy Increased sensitivity to certain substances (foods, animal products, pollen) the introduction of which into the body may cause harmful symptoms (asthma, hay fever and others).

Allopolyploid A polyploid containing chromosome sets from different species.

Allotetraploid An allopolyploid in which the sets of paternal and maternal chromosomes have been doubled in number.

Antennae A pair of jointed feelers on the head of an insect. They are concerned in perceiving scents.

Anther Structure producing pollen.

Antibody (See *Antigen*.)

Antigen A substance that can react with a related one (its *antibody*) in a specific and often dangerous way, as in the agglutination of red blood corpuscles.

Aposematic colouring Conspicuous colours indicating that a species is protected by unpleasant (distasteful) or dangerous qualities.

Autopolyploid A polyploid in which the multiplied chromosome sets are all derived from the same species.

Autosome Any chromosome other than a sex chromosome.

Autotetraploid An autopolyploid in which the chromosome sets have doubled in number.

Back-cross A mating between a heterozygote and a homozygote.

Batesian mimicry The resemblance for protective purposes of a palatable species to a distasteful or dangerous one.

Bud sport A shoot growing from a body cell in which a mutation has occurred.

Cells The units into which protoplasm is usually divided.

Character In genetics, any effect of a gene (on anatomy, physiology, mental qualities, or habits).

Chiasma An interchange of material between two of the four chromatids belonging to different but homologous chromosomes. Since they do not interchange partners, they cross over in the form of an X.

Chlorophyll The green colouring matter of plants.

Chromatids The two bodies produced by the longitudinal splitting of a chromosome when dividing. They later form two new chromosomes.

Chromatin A substance in the nucleus that can be made visible by certain stains. The chromosomes take it up when dividing.

Chromosomes Paired bodies in the nucleus of the cells. They carry the genes.

Cline A change in the structure or physiology of an organism taking place gradually over a given area of country.

Correlation. The study of simultaneous variation.

C.O.V. (See Cross-over Value.)

Crossing-over An interchange of corresponding blocks of material, and therefore of genes, between two of the four chromatids at meiosis. They are derived from different but homologous chromosomes.

Cross-over Value The classes in which recombination between two pairs of alleles has occurred, expressed as a percentage of the total offspring.

Cryptic colouring Colours that help to conceal an animal by making it resemble its surroundings.

Cytology The study of cell structure.

Cytoplasm The protoplasm of cells, other than the nucleus.

Deletion The loss of a segment from a chromosome.

Deuteranomaly Partial inability to see green.

Deuteranopia Inability to see green.

Diploid Term used of cells having two members of each chromosome pair (represented as '2n'), or of an organism composed of such cells.

DNA Deoxyriboneucleic acid. The substance of which the genes are composed.

Dominant A character that is as fully developed when the gene controlling it is heterozygous as when it is homozygous (compare *Recessive*).

Duplication An addition to a chromosome of a fragment of homologous material.

Ecology The relation of living organisms to their environment.

F1 The first filial generation: the offspring of a given cross.

F2 The second filial generation: the grandchildren of a given cross obtained by interbreeding the F1 generation.

Fertilization The union of the male and female gametes.

Gametes Reproductive cells of either sex, both in plants and animals.

Gene A hereditary unit that controls the development of definite characters.

Gene complex The system produced by the whole of the genes of an organism, interacting and with multiple effects.

Genetics The study of heredity and variation.

Genetic variation Variation produced by changes (recombinations or mutations) in the genes or chromosomes.

Genotype An organism judged by its genetic constitution (compare *Phenotype*).

Germ cells Cells set apart for the production of gametes.

Gland An organ that manufactures substances of use to the body.

Haemoglobin An iron compound, being the red respiratory pigment of vertebrates. It is carried in cells (corpuscles) floating in the blood.

Haploid Term used of cells with only one member of each type of chromosome (represented as 'n'), or of an organism composed of such cells.

Hermaphrodite An individual in which the two sexes are combined.

Heterogametic sex That with dissimilar sex chromosomes (X and Y).

Heteroploid Term used of cells that contain a chromosome too few or too many, or of an organism composed of such cells.

Heterostyly Term used of flowers in which the anthers and stigma are at different positions within the flower, so favouring cross-fertilization.

Heterozygote An individual in which the members of a given pair of genes are dissimilar.

Hibernation A quiescent state in which some animals pass the winter.

Homogametic sex That with similar sex chromosomes (X and X).

Homologous chromosomes The members of the same chromosome pair.

Homostyly Term used of flowers in which the anthers and stigma are at the same level within the flower, so favouring self-fertilization.

Homozygote An individual in which the members of a given pair of genes are similar.

Hormone A stimulatory secretion which, when discharged into the blood, affects a distant part of the body.

Imago A perfect insect after it has emerged from the chrysalis.

Immune reaction One in which the presence of an antibody is stimulated by the presence of an inappropriate antigen.

Incompatibility The failure of pollen to fertilize certain types of its own species because it does not grow on, or else does not penetrate, their stigma, owing to its genetic constitution.

Inversion A section of a chromosome that has broken off and re-attached itself in its former position, but the wrong way round.

Larva A caterpillar, or similar early stage in the development of insects.

Lepidoptera Butterflies and moths.

Lethal Term used of a gene, or genetic constitution, that kills the organism that possesses it.

Linkage The tendency for certain genes to remain together, instead of assorting independently, because they are carried in the same chromosome.

Locus The position occupied by a gene on a chromosome.

Maturation The period during which the development of the gametes is completed. It includes meiosis.

Meiosis The occurrence of two divisions of the nucleus with but one division of the chromosomes. These are the last two divisions of gamete development, and give rise to cells with one member only of each chromosome pair. Crossing-over occurs during the first meiotic division.

Melanins Black or dull-red pigments containing nitrogen.

Metabolism The life processes of the body: building up protoplasm from food, and breaking down compounds with release of energy.

Metamorphosis A sudden alteration in form, without growth: the change from a caterpillar to a chrysalis, or from the latter to a winged insect.

Mimicry The resemblance of one species to another for protective purposes.

Mitosis The process by which the nuclei of the cells normally divide. It ensures that each of the daughter nuclei receives a longitudinal half of every chromosome.

Modifying genes Genes which modify the characters produced by other genes. They may be without detectable effect themselves.

Mullerian mimicry The joint resemblance of several protected species to each other to minimize predation.

Multifactorial variation (See *Variation, multifactorial.*)

Multiple alleles Detectably different genes produced by a number of mutations at the same locus. Their effects differ in degree rather than in kind.

Mutant The gene resulting from a mutation in a stock under examination.

Mutation The inception of a heritable variation. It is of rare occurrence.

Non-disjunction   The failure of a pair of chromosomes to separate from one another at cell division.

Nucleus   A specialized part of the protoplasm within all typical cells. It is enclosed in a membrane, except when dividing; and contains the chromosomes.

Ovary   The structure in which the eggs are formed.

Ovule   The seed rudiment of a plant.

P1   The parental generation, the individuals of which are mated to produce a given cross.

Parthenogenesis   The development of an egg without fertilization.

Phenotype   An organism judged by its appearance (compare *Genotype*).

Plasma   The nearly colourless liquid of the blood in which float the corpuscles (blood cells).

Pollen   The minute structures that carry the male gametes in plants.

Polygenes   Genes having small, similar and additive effects, tending therefore to produce continuous variation. They may be widely scattered on the chromosomes.

Polymorphism   The occurrence together in the habitat of two or more discontinuous forms of a species in such proportions that even the rarest of them cannot be maintained merely by recurrent mutation.

Polyploid   Term used of cells in which the chromosome set is multiplied to higher than the diploid value, or of an organism composed of such cells.

Polysomic   Term used of otherwise diploid cells having a chromosome occurring once only or more than twice, or of an organism composed of such cells.

Prophase   The stage preparatory to cell division when the chromosomes become evident. It is then that in meiosis the chromosome pairs associate together, and their chromatids interchange blocks of material.

Protanomaly   Partial inability to see red.

Protanopia   Inability to see red.

Protoplasm   The living substance of an organism.

Pure line   The descendants of a single self-fertilized individual homozygous for all its alleles.

R2   The offspring of a back-cross.

Recessive   A character that is only expressed when the alleles controlling it are homozygous (compare *Dominant*).

RNA   Ribonucleic acid. The substance that, in its diverse forms, passes the information supplied by the genes on to the cytoplasm, where it is carried into effect.

Secondary sexual characters The organs of sex other than the 'primary' ones (those forming the gametes).

Segregation The separation from one another during meiosis of the members of the paired genes constituting the alleles. Also the result of that process, in which the offspring of a cross separate into distinct classes in definite proportions.

Sex chromosomes The X and Y chromosomes.

Sex-controlled inheritance Applied to characters that can only manifest themselves in one or the other sex. The genes responsible for them can be carried in either sex.

Sex linkage Control of characters by genes carried in the sex chromosomes.

Somatic cells The cells of the body, as apart from the germ cells.

Somatic mutation Mutation taking place in the somatic cells, instead of in those forming the gametes.

Sperm The male reproductive cell in animals.

Stamen Part of the male structure of a flower, consisting of a stalk swollen at the end to form an anther.

Stigma The surface of the style of plants. The pollen is received on it.

Structural interchange An exchange of fragments between non-homologous chromosomes.

Style A projection from the ovary of plants. It takes the form of a rod carrying the stigma at its end.

Supergene A group of two or more major genes (that is, not polygenes) responsible for different sets of characters, but so closely linked that they act as a single switch in controlling the group of alternative forms.

Testis The (generally paired) organ in which the male reproductive cells (sperm) are formed.

Tetrad The four bodies that arise when the homologous chromosomes come together in pairs during meiosis and split into chromatids.

Thorax The chest.

Translocation The attachment of a segment of one chromosome to another, non-homologous one.

Trisomic Term used of otherwise diploid cells in which one chromosome is present three times, or of an organism composed of such cells.

Variation, multifactorial Variation due to a number of genes acting cumulatively, though not necessarily as polygenes.

Viability The capacity to survive.

X chromosome A chromosome carrying genes concerned in sex determination. There are generally two X chromosomes in one sex and a single one, partnered by Y, in the other.

Y chromosome   The partner of the X chromosome in one of the two sexes. It carries few genes, and is often not directly concerned in sex determination.

Zygote   The first cell of a new individual, produced by the fusion of the gametes.

INDEX

(Human diseases are indexed separately, not under Man)